Where the Gods Reign

Plants and Peoples of the Colombian Amazon

Richard Evans Schultes

Preface by

Mark J. Plotkin
Director, Plant Conservation
World Wildlife Fund

WWF

SYNERGETIC PRESS

SP

WWF

Synergetic Press is proud to support the efforts of World Wildlife Fund to protect the world's endangered animal and plant life. A percentage of the sale of this book goes to support World Wildlife Fund.

Published by Synergetic Press, Inc.
  Post Office Box 689, Oracle, Arizona 85623
  24 Old Gloucester Street, London WC1 3AL

Cover and book design by Kathleen Dyhr.
Typesetting by Synergetic Press.

Library of Congress Catalog Card 88-90515
ISBN 0 907791 13 1

Printed in the United States of America by Arizona Lithographers.

Contents

Dedication

During my nearly half century of botanical and ethnobotanical research in the Colombian Amazonia, hundreds of individuals in governmental, educational, military and ecclesiastical circles as well as many settlers and Indians have generously and materially contributed to the success of my work. Would that I might name them all but that is beyond the realm of possibility. They must, therefore, accept my deepest thanks collectively for their friendship and help.

I feel constrained, however, to express my indebtedness to four Colombians whose preoccupation with environmental preservation has been outstandingly evident and to whom I owe a special debt of gratitude for personal encouragement. It is, therefore, my pleasure to dedicate this volume to the following gentlemen:

His Excellency, Dr. Belisario Betancur, President of the Republic of Colombia, 1982-1986, fervent conservationist and friend of the natural sciences, during whose presidency a number of major biological preserves were created, several of them in the Amazonia, and wise laws for protection of the environment were enacted through Inderena, the conservation arm of the Ministry of Agriculture.

Dr. Mariano Ospina H., Director of the Colombian Agricultural Credit Bank, 1982-1986, whose many years of staunch efforts towards preservation of natural resources led up to the establishment of the *Plan Integral de Desarrollo del Predio Putumayo*, an area of 14,800,000 acres dedicated to the exclusive use of the native indigenous population and maintenance of the ecosphere.

The late Don Rafael Wandurraga, citizen of Leticia, whose fatherly treatment of and preoccupation for the welfare of the Indians underlay his commercial relationship with the native population and whose interest in animals and plants was reflected in his willingness to befriend visiting scientists and in his life-long respect for protection of the tropical forest.

The late Don Miguel Dumit, Colombian business man, whose ardent love of the Vaupés encouraged him to establish one of the first airlines to the region and whose generous magnanimity to governmental, commercial, ecclesiastical and cultural entitites and individuals contributed greatly to the understanding of this vast, neglected and almost undisturbed tropical region and to the protection of its aboriginal peoples.

Preface

The spectacular photographs and superb text of *Where the Gods Reign* require no external explanation -- they can stand alone. Yet I feel that the reader can more fully appreciate this book if he learns a bit more about the author, Richard Evans Schultes.

What kind of man is Schultes? Below is a piece abstracted from his 1947 field journal, written while traveling up an Amazon tributary:

> The barge leaks alarmingly; an Indian boy bails continuously, the sloshing and scraping keeps us awake every night. One side of the cabin is missing, exposing us to rain. Going through the first rapids today was very difficult and dangerous -- I feared we wouldn't make it because we were balanced on a sharp rock. I am discouraged because the formaldehyde I bought is very inferior and nearly all of my collections for the past month and a half are rotten ... This afternoon I came down with a very high fever. Have rheumatic pains in every limb and back, continuous nausea, some vomiting ... Vomiting continuously, very weak, probably mostly from malnutrition -- we have had no warm food, only a tin of sardines for supper last night ... A strange accident happened this morning. About 5 a.m. there was a jolt and a loud crashing and splitting of wood. The barge had run into a leaning tree along the river's edge and the already badly damaged cabin was smashed. With my flashlight, I saw that the tree was in young fruit, with a recently fertilized ovary, that is, so I broke off a few branches to put in the press later. When dawn came, I examined the plant -- it was *Micrandra minor,* which I am especially anxious to collect! Naturally, all on board were highly amused at my joy in collecting branches from a tree that had caused them so much fright. Late in the afternoon we came upon a most wonderful sight; three distinct peaks rising above the forest canopy, the Tapir Mountains. I also have noticed beautiful coconut trees here, disproving the notion that the palm cannot thrive far from the sea. While the boat is being patched up, I am anxious to do some collecting along the bank ...(Next day) ... overdoing brought on violent vomiting with blood, am very weak. Will have to leave the intensive work until we come back downstream, when I hope to feel better.

There can be no doubt that Richard Evans Schultes is one of the great explorers of this century. After completing his Ph.D. dissertation on the ethnobotany of the Indians of Oaxaca, Mexico, Schultes decided to initiate a study of plants employed in the manufacture of arrow poisons which were then becoming very important in surgery as muscle relaxants. In 1941, he traveled to the northwest Amazon to carry out his research, and he basically remained there until 1954. Even today, the northwest Amazon remains one of the most remote regions in the world, cut off by the Andes on the west and dense jungles and impassable rapids on all other sides. Most of the Indian inhabitants had never been studied and, to this day, some have probably never been contacted. Schultes worked in this region for fourteen years, living with the Indians, participating in native dances and other rituals while carrying out ethnobotanical research. Although he eventually went to work on the USDA project to harvest natural rubber from the Amazon to aid the war effort, he continued collecting, eventually sending home over 24,000 plant specimens. Almost 2,000 of these plants are employed as medicines or poisons, while others are used in many ways -- from clothing to contraceptives.

Many of the ethnobotanists working in the tropics today are following in Schultes' footsteps. He demonstrated the importance of focusing specifically on ethnobotany rather than collecting data on useful plants as an adjunct to some other avenue of research. Schultes has always stressed the interdisciplinary approach, not only incorporating the botanical and anthropological components but also the chemical and pharmacological. Perhaps most importantly, he began writing about the importance of conserving ethnobotanical lore long before most biologists were aware of *any* conservation problems in tropical regions. In 1963, he wrote:

> Civilisation is on the march in many, if not most, primitive regions. It has long been on the advance, but its pace is now accelerated as the result of world wars, extended commercial interests, increased missionary activity, widened tourism. The rapid divorcement of primitive peoples from dependence upon their immediate environment for the necessities and amenities of life has been set in motion, and nothing will check it now. One of the first aspects of primitive culture to fall before the onslaught of civilisation is knowledge and use of plants for medicines. The rapidity of this disintegration is frightening. Our challenge is to salvage some of the native medico-botanical lore before it becomes forever entombed with the cultures that gave it birth.

Schultes' importance as a taxonomic botanist should not be overlooked. He has personally discovered many new species of plants and described many of them himself. One measure of the high regard in which he is held by his fellow botanists is to note that they have named many species *Schultesii*.

Richard Evans Schultes has also made major contributions to economic botany. He has published over twenty-five papers on the rubber tree *(Hevea)*, several important papers on palms and orchids, and major articles on *Micrandra* and *Herrania* (close relatives of rubber and chocolate, respectively), two genera whose importance to our industrialized society will increase as genetic engineering facilitates inter-generic crosses.

In addition to publishing seven books and over 380 scientific papers, Schultes is a great teacher. He inspired countless students (myself included) to pursue careers in the natural sciences. His influence on young people has by no means been restricted to his students in Cambridge -- he has traveled all over the world, giving lectures and serving on academic committees so that students who could not afford access to a Harvard education would have the benefit of his knowledge and experience.

Richard Evans Schultes is a remarkable man who has already had a major impact on several different fields of study. As the international conservation movement matures, shifting its emphasis from protecting cute animals to conserving the tropical forests which support life on Earth, Schultes' status is being changed from being merely a major influence to recognition as one of the Founding Fathers.

Where the Gods Reign appears at a particularly opportune moment when the general public is demonstrating increased interest in rainforests and tropical ecosystem conservation. This book offers the opportunity to explore the Colombian Amazon with a singularly qualified guide. The pictures and text of *Where the Gods Reign* communicate a deep understanding and great love of the peoples and plants, an understanding and appreciation developed from years of living in these rainforests.

In 1941, the noted Harvard economic botanist Oakes Ames wrote to his student Richard Evans Schultes:

> I don't know of any better subject in the botanical field on which to build
> such a personality than economic botany with its tremendously important
> human implications and its call for searching investigations among things
> that really matter. If I were a young man beginning my career all over again,
> I should try through intensive research in economic botany and ethnobotany
> to bring more light into the intellectual realm and to take my place, not in
> a laboratory cubicle, but in the world ...

Where the Gods Reign shows how well Richard Evans Schultes followed that advice.

Mark J. Plotkin, Ph.D.
World Wildlife Fund

Foreword

In putting together this book of pictures my only wish is to share with others many happy days that I spent in the forest of the Colombian Amazon. It makes no claim to complete coverage in any field, nor is it intended in any way to be a technical study. On the contrary, my hope is that I may give my readers a glimpse of some of the highlights of the region and to write about them simply, directly and briefly.

It has not been easy for me to write a popular book, partly because, as a scientist, I tend to give extreme importance to detail. The preparation of this volume has, therefore, been good discipline for me, making me concentrate on a few interesting aspects without undue preoccupation that many more of equal interest may not even be mentioned.

If I have succeeded in producing an entertainingly instructive book that imparts the peace of years spent surrounded by nature in a primeval forest, then I shall truly be satisfied; for it will then not be only the plant sciences that have benefited from my work. It is a big task. In the face of the grandeur of nature in this remote region, I have many times felt unequal to the task. Often I have set the work aside, only to take it up again later in the belief that, after all, it is something of an obligation to share with others what so few have been fortunate enough to experience themselves. Had I neglected this obligation, perhaps all through my life the pangs of conscience might have reminded me of my failure to try to give back to the world something that the world was good enough to give to me. Hesitancy that I could do the task delayed the book now for too long a time, but it may have helped me to prepare a better book than if I had boldly hastened to write it. For, in essence, one of the greatest lessons that can come from long residence in the solitude of tropical forests is that everything has its own unhurried time.

As a young botanist, armed with a bright, new doctor's degree from Harvard University, I decided in 1941 to begin a study of medicinal, narcotic and poisonous plants used by the natives of the northwestern part of the Amazon Valley. Little did I realize in 1941 that good fortune would allow me to spend an almost uninterrupted fourteen years in this remote area and that, after returning to Harvard University in 1954, I would be able to make yearly trips back to Colombia alone or with students interested in tropical botany.

During and after the war years, my principal efforts in plant exploration were directed towards an investigation of rubber trees of the genus *Hevea* which are native to these forests. My task involved immediate exploitation of wild stands for the war effort and, later, for collection of living material destined for future use of new germ plasm in the improvement and alteration of the cultivated *Hevea* tree. Since the northwest Amazon is the area of greatest diversification of *Hevea,* it was necessary to study phytogeographically the sparsely known flora of the region in order clearly to understand the evolution of this group of plants. Consequently, my travels in the northwest Amazon were planned to cover as nearly as possible the area of distribution of the rubber tree, with particular attention to rivers where rubber exhibited greater diversity or some other peculiarity that might be of interest to any genetic programme.

Most of my exploration was concentrated in that part of the Amazon Valley lying within the boundaries of the Republic of Colombia, but similar studies were carried out for lesser periods in Loreto and Madre de Dios in Peru and in the Rios Negro, Madeira and Tapajóz of Brazil. It is, however, to the Amazon basin of Colombia -- comprising one-third of the Republic and the least known part of the great Amazon Valley -- that this book is primarily devoted.

In the course of these years of travel, I took thousands of photographs. Most of them, naturally, were records of plants or of plant formations. Occasionally, however, there was a chance to photograph other aspects of nature, and of the people of the region, and this I did without any definite plan or purpose.

The possibility of preparing a book of chosen photographs had never passed through my mind; it was actually suggested to me. In 1953, the Centro Colombo-americano in Bogotá exhibited a selection of twenty-five of my photographs representative of life in the Colombian Amazon. During the exhibition, which was surprisingly popular, my friend, the late Dr. Jaime Jaramillo-Arango -- surgeon, diplomat, educator, author, man of science and letters -- strongly urged me to enlarge the exhibit into an illustrative book. The idea, at first, seemed inordinately ambitious, if not even presumptuous. Further consideration, however, and the encouragement of many of my Colombian friends convinced me that perhaps it might represent a worthwhile contribution, if it could be realized.

When I began the organization of this volume, it seemed to me that there should be some way to make it more than just another picture book. The last several decades have seen the publication of a number of superbly illustrated books on Latin American countries. Nothing of this kind, to be sure, has appeared on the northwestern Amazon, but the longer I considered the problem, the stronger became my conviction that something was needed to give my book an appeal beyond that merely of a geographical region.

The history of exploration has always interested me, and I have profited richly from my library on tropical American travel. This is a phase of history that has been rather consistently neglected in our educational systems, in our press and in our popular as well as our scientific literature. The explorer, especially the scientific explorer, has,

5

more often than not, stood at the vanguard of penetration into unknown regions and has usually been the means of acquainting the world with these remote parts of the planet. The accomplishments of many of the early explorers would, even today with modern medicines, equipment and transportation, be considered prodigious. Most of these humble and dedicated men were broadly educated and keen observers who took the time and pains to write down their experiences and what they saw and learned in the new regions. Many of them were outstanding writers -- Schomburgk, von Martius, Humboldt, Spruce, Wallace, Darwin, Bates, Belt, Crévaux, Koch-Grünberg, to name but a few -- whose pages glow with a deep and abiding appreciation of nature set forth oftentimes in truly poetic tones.

It seemed to me wholly proper that my modest pictorial contribution might, in addition to introducing its readers to a little known part of the world, try to acquaint them with brief excerpts from the writings of some of the great explorers and travellers of the past. I resolved to try to find for each of my photographs an appropriate quotation from these or similar sources. It is true that very little has been written about Amazonian Colombia -- and indeed this part of tropical America has had even today a minimum of exploration. I could, therefore, not hope to find for each illustration a quotation directly referable to the actual area. But could I not find in the wealth of this kind of literature passages -- whether from Brazil, British Guiana, Ecuador, the Orinoco or other regions of tropical America -- which at least to me expressed the *spirit* of my photographs?

As the search for such passages went on, I became more and more certain that a presentation of my photographic material and my own brief explanation of it against such an historical background not only might enhance the value of my contribution but could help me pay a personal debt of gratitude to those great travellers of the past whose pioneering spirit had, in a sense, made more meaningful the work of all modern explorers. I have, accordingly, drawn freely from a large number of writers in several languages and have translated pertinent passages in hope that my own readers will experience the thrill of discovery and the reverence for nature that possessed these men and which, through their writings, early drew my own interests to the farthest reaches of the American tropics.

There is another reason underlying my desire to acquaint readers through randomly selected photographs with the beauties and naturalness of the Colombian Amazon. The region has, even today, not suffered the rampant physical and social despoilment that has overtaken many other parts of the Amazon Valley as a result of commercial and often governmentally supported projects euphemistically called ''civilising'' and ''modernizing'' programmes. Only one of Colombia's Amazonian rivers -- the Rio Putumayo -- is navigable; all of the others are interrupted by rapids and waterfalls, some of them with several, others with many. We might say that Nature has protected the region from intrusion; boats have not been able to come west from Brazil and, except for areas along the eastern slopes of the Andes, there has been little pressure for colonists to leave the healthy, fertile mountainous parts of Colombia to wander eastward and settle in many of what to them seem to be inhospitable forests.

A number of years ago, I heard a high-ranking diplomat of another South American nation who, in addressing the Colombian senate, described the whole Amazon as "a desert of trees that had to be cleared for the benefit of mankind". Colombia, fortunately is extremely conservation-minded. Wise legislation has been passed aimed at conservation of nature and natural resources, and an admirably diverse string of "national parks" or biological preserves has been set up. It is extremely difficult to adequately police these protected areas due usually to a lack of sufficient funding, and unfortunately there are occasional cases of misuse of their land, flora and fauna. A very recent example of the nation's preoccupation with conservation has been the setting aside as a protected area some 15 million acres in the Amazon for use only by the sparse Indian population. *(Editors note: Nearly 3 million acres of which have been officially designated* Sector Schultes, *in honour of his work in the area.)*

Along with the conservation of natural resources, Colombia, through its encouragement of ethnobotanical studies, recognizes that the knowledge of the properties and uses of the native flora -- knowledge acquired over hundreds or thousands of years -- is one of the first possessions of the Indians to disappear with advancing "civilisation" -- usually in a single generation. Recent investigations by Colombian and foreign anthropologists and botanists in campaigns of what has come to be known as *ethnobotanical conservation* have recognized the detailed knowledge of an unbelievably rich flora possessed by these native inhabitants. Advancing acculturation and civilisation everywhere spells the doom of extinction of this knowledge faster even than the extinction of species themselves as a result of forest devastation.

Preservation of this acquaintance with the properties of plants of such a floristically rich and still inadequately known area can be of inestimable value to future phytochemical, pharmacological and nutritional studies by modern scientists. As has been truly stated: "The cure for cancer may come from the witch-doctor's knowledge of plants."

The photographs in this volume were taken, for the most part, in the Colombian Amazon during the 1940's and 1950's. Although changes -- and not all meritorious ones -- have come about and advanced transportation, limited growth of commercial interests and increased missionary activities have favoured these changes, the region has apparently remained remarkably similar in conditions, according to my overall view of the situation, that prevailed thirty or forty years ago. In fact, consultation with the few books of famous travellers and naturalist explorers of a century or more ago will convince the reader that, in many aspects of life in the region, little has changed.

Richard Evans Schultes, Ph.D.

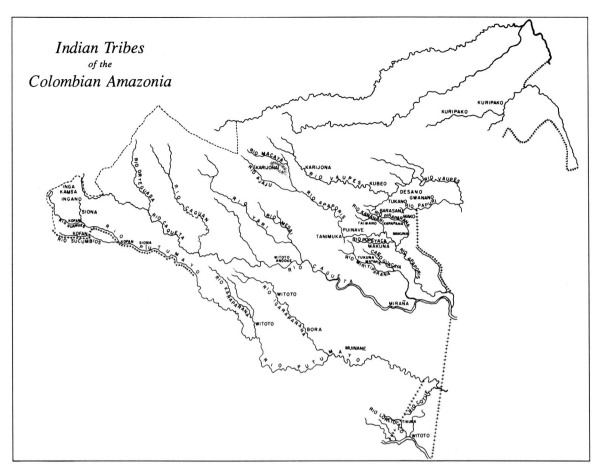

Indian Tribes
of the
Colombian Amazonia

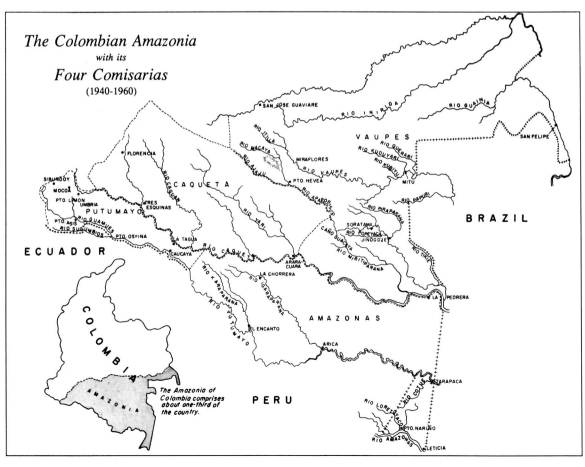

The Colombian Amazonia
with its
Four Comisarias
(1940-1960)

The Amazonia of
Colombia comprises
about one-third of
the country.

The Amazonia

The Amazonia in General

It is easier to understand the Colombian Amazonia with a brief and admittedly superficial review of the great river basin as a whole. Although different from the general Amazonia in geology, flora and inhabitants, the Colombian Amazon does form the northwestern part of the great drainage area known as the Amazonian hylea.

The Amazon basin is composed basically of metamorphic and igneous rock over which, in most areas, alluvial deposits have accumulated. The metamorphic regions are found primarily in the north associated with the so-called Guyana Shield; the igneous with the Brazilian Shield in general to the south. These two formations mingled in places in the Mesozoic era -- commonly known as the "Age of Reptiles" -- although some geologists believe that it occurred in the Cambrian period or perhaps even earlier. During the Cretaceous and Tertiary periods, when the great basin was an extensive lake, alluvial sediments were collected, often to a depth of 2,000 feet. At this period, the lake emptied out into the Pacific Ocean through a trough located in the vicinity of the present frontier between Colombia and Ecuador. Much later, in the Pliocene when the Andes were uplifted, the drainage shifted to its present eastern outlet into the Atlantic. The present form of the great basin took place very recently, during the Pleistocene or "Ice Age".

Arising at its most distant point in the high cold Andes of Peru, the Amazon River -- the water course that dominates the whole Amazon basin -- drains an area of 2,722,000 square miles, almost twice as large as that drained by any other river system in the world. The Amazon River itself, the largest river in the world in volume of water, is nearly 4,000 miles long. As the great British botanist-explorer of the last century, Richard Spruce, wrote nearly 150 years ago in a letter from the Amazon, "The greatest river in the world flows through the greatest forest".

More than 1,000 tributary rivers feed the main stream. Seven of these tributaries are 1,000 miles long, and one stretches for more than 2,000 miles. The Amazon river system drains areas in six South American countries: Bolivia, Brazil, Colombia, Ecuador, Peru and a small part of Venezuela. It empties into the Atlantic Ocean more than six million cubic feet of water per second. One-fifth of the fresh water of the world lies in the Amazon basin.

The main rivers that carry a heavy load of mud or silt are tan or yellowish and are known as "white rivers". The so-called "black rivers", flowing over sandy or granitic soil, carry little silt and have the colour of amber or strong tea. The white rivers -- the Amazon, Madeira and most of those arising in or near the Andes carry much mineral nutrient material and are relatively rich in vegetal life. The black water rivers, of which the Rio Negro and its tributaries are characteristic, are highly acidic and are, in general, poor in fishes and other animal life as compared with the white waters.

The Rio Amazon itself is sluggish, flowing slowly. It drops, for example, only 215 feet in its lower 2,000 miles -- about one and a quarter inches per mile. The mouth of the great river measures some 160 miles across. Ocean-going ships can reach Manáos, a Brazilian city more than 1,000 miles upstream; smaller ships easily navigate to the Peruvian city of Iquitos; and smaller river boats can ply some 550 miles above Iquitos. Atlantic tidal effects are strongly felt 400 miles upstream from the mouth, and ships must navigate in the lower part of the river with compasses due to the width of the river. Is there any wonder that the early explorers spoke of the Amazon as the *rio-mar,* "the ocean-river"?

There are many oxbows, lakes and great swamps. They are particularly numerous in the eastern parts of the Amazon basin, vestiges of constant movement of water and silt in this area subjected to tremendous rises and falls of water level and erosion of land. Islands of mud and silt are continually being formed, and often through erosion they disappear with amazing rapidity.

The popular notion that most of the Amazon is an immense swamp is erroneous. Much of the land -- about 80% -- lies above the annual high water levels. The higher ground naturally supports a completely different flora and fauna than the lower areas that are subject to yearly flooding. There are, in general, three different forest zones due to the influence of inundation. The so-called *terra firme* is the area that never floods; it supports a more or less open forest of very tall and usually robust trees. The *varzea* is seasonally flooded, sometimes deeply so, where the large trees often have stilt or buttress roots; and the *igapó* is completely flooded almost all year; together they comprise about 10% of the total area. The remaining approximately 10% is occupied by limited and very localized ecological zones: open grasslands called *campinas;* and especially in the Rio Negro basin isolated areas known as *caatingas,* of low and weak trees on sterile white sand.

The climate of such a vast area naturally varies, even though the Amazon Valley is totally equatorial. It is hot and humid throughout, but there is usually an appreciable difference in temperatures between day and night; a year-round average is 80 degrees Fahrenheit. Occasionally, however, several days of cool weather, especially in late June, are not uncommon. Constant high temperatures are mainly responsible for the lack of accumulation of humus, since most decomposition of organic material is, above 80 degrees Fahrenheit, the result of bacterial activity. Furthermore, high temperatures tend to prevent the accumulation of nitrogen in the soil.

Rainfall is even more important than temperature. There are three distinct climatic areas: at the mouth of the Amazon and in much of the westernmost area near the Andes, the rainfall may vary between seven and ten feet a year, with a modest difference between the wet and the dry seasons (usually from January to June); in the central parts there is a longer and somewhat more severe dry season; in the southernmost areas a stronger and longer dry season is characteristic. The difference between the ''rainy'' and the so-called ''dry'' seasons can be appreciated with the simple statistic: at Leticia, a Colombian town approximately 2,000 miles upstream, the floating dock indicates that the Amazon rises and falls some 48 feet between the two seasons, and it is often possible to paddle a canoe through the forest, sometimes even at the level of the lower-most branches of trees. The rains frequently occur in sudden cloudbursts; these strong rainfalls, unlike gentle rains, tend to wash out all soluble nutrients from the upper layers of soil.

To the uninitiated who fly over the Amazon, the forest looks like a homogeneous green carpet stretching from the Andes to the Atlantic. Nothing, however, could be further from the true state of affairs. As in tropical rain forests in general and quite unlike the usual condition in temperate forests, there is no ''dominant species'': an area of 1,900 square feet of forest in the central part of the Amazon was found to have 1,600 species of plants. One study of several forest plots indicated that from 20 to 46 different species of trees (counting trees only of four inches or more in diameter) were found per acre. This study was carried out in the eastern Amazon; in the western areas, the diversity is undoubtedly even greater.

Diversity is the rule. The flora of the Amazon hylea is estimated to have up to 80,000 species of higher plants, more than 4,000 of which are trees. The fungi, known to be extremely numerous in the wet tropics, constitute an almost virgin field of investigation and do not enter into the estimate.

How did such luxuriant vegetation evolve? The environment of the Amazonian hylea is a most complex system with species of plants, animals, soils and Indians interacting with a precision so delicate that any interruption can very easily upset the balance of nature.

The Amazonian soils are not rich. There are two principal kinds of soil. One, known as latosol or lateritic gravel, is found in extensive areas throughout the basin; it is argillaceous or clayey and is rich in iron but very poor in nutrient materials. The vegetation on the latosols tends to be forests located above the level of annual flooding. The other predominant soil is known as the podzol type; it is scattered extensively throughout the basin but is very frequent in the northwestern area of the Rio Negro and its main affluents. In this region, it is found in isolated areas of bleached white sand -- formations of small, weak trees with limited canopy cover and known as *caatingas,* a term meaning in one of the Indian languages "white sand".

The luxuriance of the forest cover of tall trees has given non-specialists the impression of an extremely fertile soil that can be the basis of agriculture as intensive and mechanised as that practised in temperate zones. Although such an impression may be enticingly convincing to non-technical observers and governmental decision-makers, it is completely erroneous and is based on no scientific evidence. It has led to tremendously fatal devastation of enormous areas, particularly in the easily penetrated regions of the Amazon -- devastation condoned and sometimes encouraged by governments that until recently have lacked the benefit of the latest scientific evaluations.

In most parts of the basin, when the forest is disturbed or exterminated, the soils are depleted in several years, sometimes in a single year. Poverty of the soil and depletion of the limited nutrient constituents has forced societies in the Amazon and many other humid tropics to develop cassava or the tapioca plant as their major carbohydrate food: this plant has a deep root that can penetrate up to four feet below the surface, since crops with superficial roots such as maize, rice and other cereals cannot be grown over any lengthy period. Successful agriculture in most parts of the Amazon will be primarily root-crop cultivation, and cultivation practised on a small scale with the so-called "slash and burn technique". Poverty of the soil, furthermore, is the basic reason why extensive concentrations of large human populations are doomed to failure if not supported by major and costly importation of foods from outside.

The Amazon region has given the world some of its most valuable economic plants, including the Pará rubber tree, cacao (source of chocolate), pineapple, the tapioca or cassava plant, the Brazil nut tree, timbó (a source of biodegradable insecticide, rotenone), achiote (a red colouring agent for foods), coca (source of cocaine) and probably the nutritious peach palm. There are sundry other species promising as crop domesticates, including several oil-rich trees and plants productive of fine timber, sources of gums, resins and waxes, as well as a great number of unstudied species that are potential source of many chemical compounds new to science, some of which might have medicinal or industrial uses.

It is true that our study of the economic value of the Amazon flora has only just begun. A noted Brazilian phytochemist specializing in tropical plants has stated the 90% of the species of the Amazon forests have not yet been subjected even to superficial chemical analysis.

When Indians fell a small area for their own use, the climax forest only very slowly regenerates and takes over. Even the Indian often has to change his dwelling-site because of soil depletion, proliferation of the leaf-cutting ant and other reasons.

Within 100 or 120 years, the climax forest, through a series of floristic successions, can re-establish itself but only in such small denuded areas. When vast expanses -- not infrequently up to one million acres at a time -- are subjected to devastation with bulldozers and other modern mechanical methods, primarily to put in grass for cattle, the climax forest will never be able to re-establish itself. In most Amazonian soils, grasses will not thrive longer than several years. Then an adjacent one million acres is felled to make more pasturage. The original area is abandoned, to be taken over by worthless scrub vegetation -- useless to man and nature. With this devastation disappear many species of plant and animals that may in the future be of extraordinary value of themselves or as tools in genetic programmes for the benefit of all mankind. The most unfortunate aspect of this whole sorry situation is that many plants are becoming extinct even before they can be described and named by botanists or chemically studied.

I once heard a high government official from a neighbouring Latin American nation address the Senate of the Republic of Colombia. Referring to the Amazon forest, he said, "We Amazonian countries must collaborate to do away with this desert of trees."

Each month in the eastern Amazon there is flattened a forested area as large as the American state of Rhode Island. And this represents only one part of the vast Amazon. Each year, around the world, an area of tropical rain forests the size of the state of California is destroyed. Neither man or the planet can much longer sustain such devastation.

No true conservationist can argue that a vast region like the Amazon should be wholly closed to man and unavailable for human use -- to be kept as a living museum. Man is, after all, a part of the ecosphere and an important part, and he must live. And if -- as it now appears -- no control of the vertiginous population growth can be achieved, the human race must have "Lebensraum". But it is absolutely necessary for the future of mankind that representative areas be set aside and preserved with their original floras and faunas -- areas of forest protected from wanton colonisation and consequent disturbance. This need is urgent not only in tropical regions but equally so in temperate zones. Felling of natural plant covers and uncontrolled exploitation of forest must be controlled and permitted only in restricted sectors and after adequate study by competent specialists in the several relevant fields.

The Republic of Colombia has an admirable record in setting aside many protected areas (called "parques nacionales") in the Amazonia and throughout the nation. Ecuador and Peru are initiating similarly wise restrictions and projects. There is yet -- even in these forward-looking nations -- much to do, but recognition of the absolute necessity of doing something for the protection of our natural, non-renewable resources is becoming widely and well established and is now leading to definite action in many tropical areas of the world.

Fortunately, the urgent message of conservation is getting across to the public, to commercial and industrial entities and to many national and international organizations. This welcome break-through is in great part due to the intensive and growing educational programmes of numerous conservationist groups working locally or on a world wide scale to point out the manifold dangers inherent and inescapable that continued devastation of natural non-renewable resources such as that which is still rampant in much of the Amazonian basin will cause to man's very survival. It has truly been said, "Extinction is forever."

The Colombian Amazonia

Although it is usually not well recognized, nearly one-third of the land surface of the Republic of Colombia is Amazonian. The waters of this part of the country find their way eventually to the Amazon and through that international waterway to the Atlantic Ocean.

The Colombian part of the Amazon is, in many respects, very different from the rest of the great basin. It is a complex and varied region. It drops from 12,000 feet on the eastern slopes of the Andes to forested plains as low as 300 feet but which are spotted with isolated quartzitic and granitic mountains, some as high as 3,000 feet. Furthermore, all but one of Colombia's Amazonian tributaries are incapable of navigation because of rapids and waterfalls. These natural obstacles have helped reduce penetration by boats from Brazil and, with a few exceptions, settlers in great numbers have not left the fertile and healthy Andean highlands to colonize the jungle areas. As a result, the Colombian Amazon remains one of the vegetatively least disturbed and anthropologically one of the most traditionally intact parts of the entire basin. According to the most recent census, the average density of population is approximately 0.29 inhabitants per square mile (0.75 per square kilometer), with much of the population consisting of Indians.

When the photographs comprising this volume were taken (1941-1961), Amazonian Colombia was administratively divided into four *comisarías:* the Putumayo, Caquetá, Amazonas and Vaupés. Recently, two new *comisarías* have been created: The Guaviare and Guainía. These two additional units have been carved out of the former Vaupés which used to be the most extensive administrative area of the nation.

The four *comisarías* -- Putumayo, Amazonas, Caquetá and Vaupés, i.e., the Colombian Amazonia, totalled approximately 155,809 square miles (403,650 square kilometers). There are few relatively large centres of population in these *comisarías:*

in the Putumayo -- Sibundoy, Mocoa and Puerto Asís; Amazonas -- Puerto Leguízamo, Leticia; Caquetá -- Florencia, La Pedrera; Vaupés -- Miraflores, Mitú.

The Colombian Amazonia has probably the richest flora of the entire hylea, due primarily to its complex geological structure. The floras of the western highland slopes in the Comisarías del Putumayo and Caquetá are typically Andean with no relationship to the vegetation of the lower regions; but since the rains of these mountain slopes flow into the Amazon watershed it is advisable to consider the area as part of the Amazon.

The various floras of the lowland forests are so extensive that anything near a complete description is not possible in such a short introduction.

In the Comisarías of the Putumayo, Amazonas and part of the Caquetá, the tallest trees of the forests are commonly members of the legume family (Leguminosae) and the spurge family (Euphorbiaceae, especially one of the important native rubber trees, *Hevea guianensis*). An apocynaceous tree of the dogbane family, known as juansoco (Couma), is locally numerous and has been exploited as a source of a chewing gum. There are numerous tall palms, especially seje (Jessenia), source of an edible oil, and mirití (Mauritia), a tree of many local Indian uses; both grow in dense patches -- the former in non-flooded forests, the latter in swamps. The diversity of species of lower under-storey trees and shrubs is bewildering, including, amongst the most conspicuous, members of the melastome family (Melastomataceae), relatives of the cacao tree (Theobroma), many species of the mulberry family (Moraceae), the Brazil-nut family (Lecythidaceae, especially species of Eschweilera and Gustavia), numerous members of the nutmeg family (Myristicaceae, particularly species of Virola and Iryanthera), a wide representation of the myrtle family (Myrtaceae, notably species of Eugenia), and many species of Palicourea, Psychotria and Faramea of the madder or coffee family (Rubiaceae). Species of Piper of the pepper family (Piperaceae) are extremely abundant. Characteristic of the forests of this area is the profusion and variety of epiphytes* made up of ferns, bromels, species of Carludovica of the Panama-hat family (Cyclanthaceae), the aroids (Araceae), and representative of the orchids and lesser known families. Likewise, characteristic of this forest is the number of extensive lianas or woody vines that belong to several groups: the bignonia family (Bignoniaceae), the soap-berry family (Sapindaceae), the malpighia family (Malpighiaceae), the moonseed family (Menispermaceae), the logania family (Loganiaceae). Members of the last two families are sources of curares or arrow poisons prepared by the Indians. Composites and grasses are noticeably rare, usually absent except as weeds in abandoned house sites. The numerous islands of the Rio Putumayo and most of the Rio Caquetá, deeply flooded during the rainy

*An epiphyte is a plant which grows upon another plant but is neither parasitic on it nor rooted in the soil.

15

season, are clothed with trees associated with the varzea vegetation: guarumo and higuerón or wild fig (Cecropia and Ficus of the mulberry or fig family, Moraceae, respectively), vara santa (Triplaris of the Polygonaceae or buckwheat family) and numerous palms such as assaí (Euterpe) and tucumá (Astrocaryum).

The major rivers in the Colombian Amazonia are the Putumayo, Caquetá, Apaporis, Vaupés, Guainía (Rio Negro) and a very short stretch of the Rio Amazon itself.

The Putumayo is formed primarily by the Rios Sucumbios and Guamués. The Guamués arises in the large lake, Laguna de la Cocha, located at approximately 10,500 feet in the easternmost part of the Departamento de Nariño. Forming part of the frontier with Ecuador and most of the border with Peru, the Putumayo is the only navigable major river the Colombian Amazonia. Relatively large boats can travel up to Puerto Leguízamo or Caucaya, where the Colombian navy has a naval base with a gun-boat. The Putumayo continues into Brazil, where it is known as the Rio Içá and joins the Amazon at San Antonio do Içá.

The Putumayo receives from the north in the middle of the its Colombian course two large tributaries, the Karaparaná and the Igaraparaná. The names mean respectively ''river of canoes'' and ''river without canoes'', indicating that the former river is navigable, while the latter has rapids. The Witoto and Bora Indians inhabit these rivers and the area between them. It is a region of unsavoury history as the main centre of the operation of the nefarious Casa Arana, the Peruvian rubber-producing company that in the earlier decades of this century enslaved, mistreated and decimated the defenseless Indian population -- an epoch that internationally became known as the ''Putumayo scandals''. The two major Indian settlements -- El Encanto on the Karaparaná and La Chorrera on the Igaraparaná -- were centres of the Casa Arana activities. The Colombian government, through its Agricultural Credit Bank, in 1986 set aside a base area of nearly 15 million acres in this region for the benefit of the Indian inhabitants and conservation of the ecosphere; it is known as El Prédio Putumayo.

Colombia possesses only a very short length of the Rio Amazonas -- approximately 65 miles of the southern banks of the so-called Trapécio Amazónico, from Leticia upstream to the Rio Atacuari. This stretch has both low and high banks and numerous large islands. Like the Amazon in general, its banks are usually subject to intense erosion when the rains swell the river and it flows with unusual force. The riverside vegetation is typical of that along the upper Rio Amazonas with many tall silk-cotton trees *(Ceiba pentandra),* the majestic palo mulato *(Calycophyllum Spruceanum* of the Rubiaceae), and a corpulent species of Erythrina. Particularly noteworthy concerning the flora of this Colombian stretch of the Rio Amazonas and its several affluent streams, the Rios Amacayacu, Loretoyacu and Boiauasú, is the presence in great abundance of the commercially most important rubber tree, *Hevea brasiliensis* -- the only Colombian territory where this valuable species occurs in natural stands.

The Rio Caquetá is an interesting river. It flows serenely for some 580 miles of its Colombian length through mostly low banks, but it encounters rocky interruptions that cause rapids and waterfalls in two localities. At Araracuara, above the mouth of the Rio Yarí, it flows through a long series of fierce rapids and waterfalls. In the 1830's, the famous German botanist, von Martius, made important plant collections along the Caquetá but was unable to proceed upstream from Araracuara when he was faced with such an indomitable obstacle. Many years later, the French explorer Crévaux was so deeply impressed with the waterfall that he published a beautiful drawing of it. The other rapids in the Caquetá are encountered above the town of La Pedrera at the Angostura de Córdoba. This rapid is due to the proximitiy of the mountain now known as Cerro de La Pedrera but classically called Cerro Kupatí. This isolated mountain, a repository of numerous endemic plants, was formerly sacred to the Indians, and the rocks in the river below the rapids are intricately decorated with complicated engravings elaborated hundreds of years ago by natives in connection with religious and mythological beliefs.

The Rio Caquetá has several large tributaries. In its western part, the Orteguaza and Caguán are relatively well populated with settlers and farmers who have come in from the higher areas of Colombia. The Rio Yarí with its affluent, the Mesaí, both with innumerable rapids, flow over a vast, uninhabited region of flat, sandstone mountains situated between the Caquetá and the Rio Apaporis to the north; the Yarí empties into the Caquetá immediately below Araracuara. Another northern tributary is the Rio Miritiparaná, free of rapids and home of the Yukuna and Matapie Indians; it enters the Caquetá above the rapids at Córdoba. On its southern side, the Caquetá receives the Rio Cahuinarí, a very long river traversing dense forests and only very sparsely and locally inhabited by Bora and Witoto Indians.

The next two major rivers north of the Caquetá are the Rios Apaporis and Vaupés, both rather similar in flora and in their concentrations of rapids, waterfalls and small mountains.

The Rio Apaporis is almost unpopulated and is poorly known. Its 1,350 miles of "black waters" flow mainly through relatively high banks. Its rapids and waterfalls, unspoiled by human hands, are natural beauties. The Apaporis is formed by two smaller rivers -- the Rios Macaya and Ajaju. They skirt two sandstone mountains: Cerro Chiribiquete, a long, flat mountain at the confluence of these rivers and the heavily eroded Cerro de La Campana, upstream in the Ajaju. Below the confluence of the Ajaju and Macaya, the Apaporis bathes a long mountain on its southern bank. This elevation is known as Cerro del Castillo because of its grotesquely erosive face. These isolated metamorphosed sandstone mountains scattered along the Apaporis and Vaupés are remnants of a once continuous chain of the ancient highlands of the

Guyana Shield. They are repositories of high endemism with floras unrelated to the vegetation of the surrounding forests but closely allied to the plants of the much higher tepuís of Venezuela and British Guiana -- mountains such as Cerro Duida, Ayuantepuí and Roraima.

The vegetation of the sandstone mountains and the associated savannahs is extremely xerophytic -- adapted to a region of high rainfall but without soil to hold the water. Almost all trees are small or even shrubby and have usually leathery leaves: very noticeable is a diminutive Bombax and a dwarf variety of the rubber tree genus, Hevea; members of the dogbane family and the melastome family are very numerous, as are representatives of the shrubby Gongylolepis, one of the few composites of the Amazon. There are also numerous representatives of the peculiar shrub Vellozia and many species of the yellow-eyed grass family (Xyridaceae), numerous unusual sedges and rare species of the rapateaceous family.

A series of small rapids are encountered in the Apaporis until, below the mouth of the Rio Kananarí, the river is interrupted by the glorious waterfalls, the Raudal de Jirijirimo. The whole river, perhaps half a mile wide, narrows to about 150 feet and then falls 100 feet to continue through a narrow chasm some eight miles long and 50-60 feet wide, with high rock walls on both sides. The river even disappears from view in a short tunnel before emerging and finally widening into a placid river again. The falls at Jirijirimo are caused by the rocky end of the neighbouring Cerro Isibukuri, a long, flat quartzitic mountain running north-south and bathed by the Rio Kananarí, home of the Taiwano Indians.

At Jirijirimo, there is an extensive and picturesque area of low forest and savannah on white sand; it is highly endemic and of great biological interest. The Colombian government intends to recognize this most unique and beautiful natural jewel by creating it a protected area.

Below Jirijirimo, the Apaporis has a number of rapids and falls, a fascinatingly beautiful one of which is the horse-shoe shaped Yayacopi between Jirijirimo and the mouth of the Rio Piraparaná. All of these tremendous rapids and falls figure in local Indian mythology. It is not difficult to understand how such frightening turbulence of normally placid waters can be awesome to the natives and, as a consequence, objects of wonder.

The major tributary flowing into the lower Apaporis from the north is the Rio Piraparaná. It has many rapids and is the home of numerous tribes, including Makuna, Barasana, Taiwano and the semi-nomadic Makú.

The Rio Vaupés -- formed by the junction of the Rios Itilla and Unilla -- is another "black water" river with normally high banks. It flows relatively slowly and is placid in most of its upper course. The major waterfall of the upper Vaupés is Yuruparí, a gracefully picturesque horse-shoe shaped falls of some 40 feet; it has entered importantly into the stories and mythological beliefs of the Kubeo Indians. In the Rio Vaupés below Yuruparí, there continues a series of treacherous rapids and smaller waterfalls. The Vaupés flows in a generally easterly direction, entering the Rio Negro in Brazil. Part of its lower course in Colombia forms the Colombo-Brazilian boundary. The Rio Vaupes was the centre of the resuscitation of the rubber producing industry during and following the Second World War.

The forest flora of the Vaupés is somewhat different from that of the other parts of the Colombian Amazon. The canopy trees are greater in variety of species: there are several species of the rubber-yielding genus Hevea and the closely related Micrandra of the spurge family. Beautifully flowering trees of Vochysia (Vochysiaceae) are numerous in species and abundant in individuals. There is a rich assortment of palms and of the arrow-root family (Marantaceae). The legume family is exceedingly well represented. The under-storey is rich in small trees, especially of the madder family: members of the genera Psychotria, Palicourea, Faramea and Duroia. Aroids, both epiphytic and terrestrial, abound, and epiphytic cyclanthaceous species are conspicuous. Along the river banks, trees of Caryocar of the Caryocaraceae, sources of edible nuts and a fish poison, are conspicuous. Myrmecophilous shrubs of the leguminous Tachigalia are found in great profusion in the riverside vegetation, along with trees of wild figs (Ficus), latex-rich representatives of Malouetia of the dogbane family and members of the Sapotaceae or sapodilla family. The rare blue-flowered orchid, *Aganisia cyanea,* is not uncommon, hidden in the riverine tangle of vegetation.

Below the settlement of Mitú, capital of the Comisaría del Vaupés, the river flows through an endless turmoil of rapids. At Javarité, the river enters Brazil, eventually to join the immense Rio Negro on its way to the Amazon.

In this lower Vaupés region, the scattered mountains are no longer quartzitic but granitic, associated with the Brazilian Shield. These mountains support a biologically less interesting and much poorer flora than that of the sandstone mountains: it is composed mostly of introduced weedy types, with few endemics. These granitic mountains have a vegetation interspersed conspicuously with usually widespread species, including grasses, bamboos, and sedges, members of the pineapple family, especially a large Pitcairnea and a few orchids, particularly dense stands of showy representatives of Epistephium.

In the easternmost part of the Vaupés, there are isolated caatingas -- areas of sparse but highly endemic vegetation of low trees and bushes growing on sterile white sand. The sand is usually densely covered with clumps of Cladonia. Conspicuous amongst the bushes and trees are several species of the genus of rubber trees, Hevea, *H.*

pauciflora var. *coriacea* and especially the strict endemic *H. rigidifolia*. A leathery-leaved relative of the rubber tree, *Micrandra Sprucei,* is found in all of these caatingas. A curious, thick-leaved, shrubby species of Pagamea and several species of Retiniphyllum, both of the madder family, form a prominent part of the vegetation. Large terrestrial aroids, particularly species of Philodendron, are common, and numerous epiphytic bromels are abundant.

The two famous botanical experts on the Brazilian Amazonia, Drs. Adolfo Ducke and George A. Black, wrote in their "Phytogeographical Notes on the Brazilian Amazon" (Anais Acad. Bras. Ciênc. 25 (1953) 1-46): "This part of the hylaea [the northwest Amazon], as well as the partly mountainous zone of transition to the flora of the Middle Orinoco basin, belongs entirely to Colombia and may perhaps be the least studied but also the most interesting part of the whole region." This statement is without a doubt very true, but we are now able to expand our concept and venture the suggestion that the Colombian Amazonia represents probably the most varied and the species-richest region of the Amazon basin.

The Indians of the Colombian Amazonia

The northwestern part of Amazonia -- especially the area lying within the Republic of Colombia -- is the home of a large number of Indian tribes. These tribes, together with their jungle world and the plants that they know and use from their ambient vegetation, are the subject of this volume of photographs.

These numerous tribes speak a mosaic of different languages belonging to several wholly unrelated linguistic families. The Tukanoan speakers divide into two centres: an eastern group including the Tukano, Gwanano, Taiwano, Kubeo, Karapaná, Desana, Barasana and Makuna; and a western group, the Siona and Koreguaje. Between these two centres there are several Witotoan groups: the Bora, Miraña, Muinane, Andoke and Witoto proper. There are also in this area between the two Tukanoan groups the Arawakan Yukuna, Tánimuka, Matapie and Kawiyarí; other Arawakan groups -- the Kuripako and Baniwa -- live further to the north on or near the Rio Guainía. In this same area there are the Puinave who speak a language related to that of the Makú; these Makú are small semi-nomadic groups living interspersed with the eastern Tukanoans, while the Puinave have migrated in small numbers into the Vaupés, living now along the Rios Vaupés and Apaporis. The Karib family is represented by the Karijona, now living in the uppermost Rio Vaupés with a few near La Pedrera on the Rio Caquetá; and further to the south, in the region of Leticia and adjacent parts of Brazil, there are Tikuna whose language may also be of Arawakan stock. Higher up, where the forest meets the foothills of the Andes, there dwell groups of Kofán and Kamsá, both of uncertain linguistic affiliation, as well as the

Ingano and Inga tribes who speak a Ketchwan language related to the speech of the Incas.

Living in a floristically rich environment, most of the tribes of the Colombian Amazonia have been exposed to contact with outsiders seeking dye-woods, medicinal plants, drugs, rubber and balata, animal skins and gold. Though these forest-based industries have relied on Indian knowledge, skills and labour, they have often subjected the Indians to conditions little short of slavery and caused many deaths. In some cases the tribes have maintained their numbers and even have an expanding population; in others, populations are limited or in danger of disappearing altogether.

Besides the extractive industries, religious missionaries have been a further agent of cultural change. The missions have often protected the Indians from exploitation and brought much needed medical aid to supplement indigenous knowledge, but they have also worked hard to change many of the customs and traditions of the sundry Indian groups. In particular they have battled against the influence of the shamans or medicine men, the repositories of tribal lore and usually men with an impressive knowledge of the properties and uses of plants. A limited amount of this knowledge is already known to science; much awaits recognition and recording by ethnobotanists working in the field; and unfortunately some is already lost. Today some of these tribes live in close contact with non-Indian neighbours, have adapted to them, and have adopted many of their habits and beliefs -- though almost always only as a superficial overlay.

The Witoto and Bora were almost unknown until the days of the rubber boom at the end of the 19th and early 20th centuries. In the early part of this century, the infamous and brutal treatment (virtual slavery, malnutrition, overwork, torture and wanton murder) especially of these Indians by the Casa Arana -- a Peruvian rubber company then in possession of the Colombian Comisaría del Amazonas -- is reported to have reduced the population of these and related tribes from some 50,000 to 7,000 individuals during a 10-year period ending in 1920 when the British and Dutch colonies in Asia were able to fill the world's demands for cheaper and better rubber from plantations. The establishment of these plantations in the Far East redounded to the benefit of the Indians, since the extraction of rubber from the wild could not compete with plantation production, thus saving thousands of lives of Amazon natives who were freed from merciless exploitation when the forest industry all but died out.

Although the Witoto and Bora are now highly acculturated, many old customs persist. Some still live in multifamily houses, fish with nets and fish poisons and hunt with traps, use coca and ''drink'' tobacco. The shotgun has displaced the blowgun and curare. They still celebrate their tribal ceremonies, even though most have long been exposed to Christianity. Shamanism is still strong amongst the Bora and Witoto.

The shamans, who have a wide knowledge of herbs which they use, call on birds and mammals, especially the jaguar, for help. They use tobacco snuff. Most Witoto do not use ayahuasca, but a group living at Araracuara on the Rio Caquetá smoke the dried leaves instead of drinking a decoction of the bark. Witotos and Boras eat pills made of the hallucinogenic ''resin'' of the Virola tree but, unlike many other Indians, do not employ it as a snuff. Their acquaintance with the properties of plants that they use medicinally is very extensive.

The Tikuna live along the upper Rio Amazonas and its tributaries in Brazil and in the Colombian Trapécio Amazónico. They were first mentioned in 1641 in missionary reports, and this contact has led to the loss of much of their traditional culture. A peaceful people living mainly on fishing, cassava, yams and wild tubers, they are now highly assimilated to the peasant culture of settlers in the region. Their location along the main river has meant a long and continued history of contact with outsiders. They are famous for the ceremony that marks a girl's first menstruation. During this puberty ritual, still practised, the girl has her hair pulled out by the elder women, and the ceremony involves music, bathing and the use of elaborate bark cloth masks. Their painted bark cloth, along with basketry and ceramics, now form important items of trade. The Tikuna are still much devoted to shamanism and are especially apprehensive about the hexing powers of aged women and certain trees that cause disease by kidnapping children's souls. They have many medicine men for treating disease which they diagnose through the chewing of tobacco; their knowledge of the medicinally important plants is extensive, but this knowledge appears to be shared with most of the general population. Much of their complex religion and mythology has now been obliterated by a superficial overlay of Christianity.

The Kuripako live primarily along the Rio Guainía, headwaters of the Rio Negro; they are distributed on both the Colombian and Venezuelan sides of the river. While their use of medicinal plants is extensive, knowledge of their properties seems not to be limited to the shaman, as in most tribes, but to be rather generally known by the population. Despite long contact with missionaries and other western influences, the Kuripako have maintained much of their traditional culture. They still make, for example, the great wooden boards with quartz chips set in tree resin which all other groups buy to grate their cassava. They also trade beautiful pottery with designs painted with a red pigment extracted from the leaves of a vine and glazed with tree sap.

The Yukunas and Matapies inhabit the headwaters of the Rio Miritiparaná. In the 1920's and earlier, they were deeply involved in the production of balata, a non-elastic type of rubber from several species of Amazonian trees and formerly employed in making storage batteries and under-sea cables. They hold an elaborate ceremonial dance, the Kai-ya-ree, to celebrate the annual ripening of the peach palm *(Guilielma speciosa)* in April. Dressed in great masks of bark cloth with faces made from resin of various trees, they represent their relationships to different animals through songs, dances and music. This dance lasts usually three to four days. Preparations for the ceremony are extraordinarily extensive, involving hunting for and smoking of meat, manufacture of great quantities of fariña (meal from the peach palm fruit) and coca powder. They have a very extensive knowledge of medicinal plants and customarily chew large amounts of coca cultivated in gardens which also contain cassava and many fruits. They also cultivate many very sweet varieties of pineapples and use them in preparing a fermented drink, chicha.

The Matapies comprise only a few individuals closely related to and living together with the Yukunas. Members of these two tribes are robust and strong physical types. They also chew excessive amounts of coca and cultivate great extensions of the coca plant and cassava, together with a surprisingly complete selection of fruit plants.

Makú is a name applied to the many small groups of semi-nomadic Indians living deep in the forest away from the major rivers and often isolated from one another. The Makú of the Rio Piraparaná were studied by Dr. Peter Silverwood-Cope. They live mainly on hunting and fishing and eat many wild tubers and fruits gather in the forest. They hunt with blow-guns and prepare powerful curares for blow-gun darts and poisoned arrows. They practise only limited agriculture and get cassava and other cultivated foods from their Tukanoan neighbours in exchange for curare, vine baskets and smoked meat. The Tukanoans of the region consider the Makú to be their serfs or servants who will do sporadic work for them before disappearing back to the forest. They have no canoes and cross streams by wading or swimming. In the last century, the Tukanoans sold many Makú slaves to Brazilian colonists, and today their numbers are greatly reduced.

The Kofán live along the Rio Sucumbios in the Colombian Comisaría del Putumayo and in adjacent parts of the Oriente of Ecuador along the Rio Aguarico. They were first contacted in 1536 by Catholic missionaries. Later they were enslaved by the Spaniards and their number greatly decreased. In the 1940's the size of their population in Colombia was estimated to be under 300.

The discovery of oil in the regions inhabited by the Kofán has led to severe disruption of their traditional life, and road-building has opened up their lands to colonisation. Their language has not yet been definitely assigned to any large linguistic group, although the suggestion that it is a Chibchan language has been offered; it may also be the sole survivor of a separate linguistic group. The Kofán, at least until recently, excelled in the preparation of many kinds of curare and had arrow-poisons made from a large number of plants not previously known as sources of poisons and consequently not used by other tribes of the northwest Amazon. They possess perhaps the most extensive plant pharmacopoeia of all tribes in the Colombian Amazon. They chew coca and may occasionally use ayahuasca. The Kofán are friendly, cooperative and loyal and maintain close family ties.

In the highland Valley of Sibundoy there are two tribes -- the Inga and the Kamsá. The Inga spread to the region of Mocoa in the warmer lowlands where they are known as the Ingano. Both tribes have a long history of resistance to outside pressures, adapting themselves to new conditions and striving to preserve their ancient culture. They still dress in the *cusma,* a kind of cotton cloak, and wear pounds of glass beads around the neck. The language of the Kamsá has not yet been classified with certainty, but it is thought to be Chibchan. There are some 2,000 Kamsá, all living in the valley at 8,000 feet, and they are thoroughly familiar with the Andean flora on the surrounding mountains. Shamanism is still practised and plays a very important role in their lives. Their medicine men cultivate a variety of medicinal plants, including many hallucinogenic species such as ayahuasca vines and Brugmansia shrubs with large red or white trumpet-shaped flowers, and even import ayahuasca from the warmer lowland areas. They are highly skilled in their profession and are consulted by whites and Indians alike. Famous throughout Colombia, they can often be seen selling herbs in city markets, and they also travel abroad to other Andean countries.

The Puinave are, for the most part, a people who live in the Orinoco basin of Colombia -- the so-called Llanos. Their concentration is along the Rio Inirida, but many have migrated into severals parts of the Vaupés, especially in the days of rubber-collecting during the war emergency in the 1940's. Originally they were farmers and were well versed in the use of medicinal and toxic plants; they formerly prepared curare and at present are acquainted with numerous fish poisons. They still use snuffs prepared from Virola (yakee) and Anadenanthera (yopo) in their diagnostic and shamanistic rituals. A comparatively high cultural type, the Puinave are excellent manufacturers of baskets and hammocks.

In the northwestern Amazon, there are two centres of Tukanoan Indians: one in the area of the Rio Vaupés and its tributaries straddling the equator on the frontier between Colombia and Brazil and the other further west in the area of Rio Putumayo. Though they speak similar languages, the two groups are culturally separated and have apparently been isolated for a long time.

The eastern Tukanoans comprise the major concentration of this linguistic group. There are some 60 sites along the Rio Vaupés from Mitú in Colombia to Ipanoré in Brazil and on the Rios Papurí, Tikié and Curicuriarí.

The eastern Tukanoans number some 2,400 overall but are divided into a number of small tribes: the Desano and Tukano live on the Rios Papurí and Tikíe, the Piratapuyo on the middle Rio Vaupés and lower Papurí. Neighbours of the Gwanano, the Kubeo are on the middle Vaupés and its tributaries, the Kerarí, Kuduyarí and Kubiyú. Further south, on the Rio Piraparaná, a tributary of the Apaporis, live the Taiwano, Barasana and Makuna. On the headwaters between the Rios Piraparaná and Papurí are groups of Karapaná, and a few more Karapaná live on the Vaupés below the falls of Yurupari.

Traditionally these peoples lived in large communal houses or *malocas,* but now many are located in villages built around mission stations. Several attempts to establish missions -- from 1852 to 1880 -- failed, but from 1914 onwards, missionaries have come to play an increasingly dominant role in the Indians' lives.

Plant medicines and poisons play an important role in the lives of the eastern Tukanoans and are often employed in addition to their shamanic curing rituals. The first botanical work on caapi, the hallucinogen of the genus Banisteriopsis, was done in their area and the specific name *caapi* comes from a Tukanoan language. Coca is widely used, especially by groups like the Barasana and Makuna who chew the powdered leaves mixed with leaf-ash.

The Kubeo are renowned for their knowledge of poisons and also for their bark cloth masks, similar to those of the Yukuna but used in a funeral ceremony to represent the animals who come to welcome the spirit of the deceased. The Kubeo suffered relatively harsh treatment during the rubber boom in the first two decades of the present century, and they have also been strongly influenced by evangelical Christianity. But they are a very adaptable people, a feature shown in their readiness to adopt new crops such as maize, rice, taro, etc., and in the fact that they have recently begun to revive many of their old customs. In the 1940's, their population, spread in numerous settlements but concentrated along the Rio Kuduyarí, was calculated to be some 2,200. Like other Vaupés tribes, the Kubeo are excellent farmers who grow a variety of plants for food, medicines, poisons, fibres and pigments. Their basic diet comes from cassava, from which they prepare fariña, a flour that is their staple carbohydrate food and which they sell to outsiders. Taro, yams, bananas, plantains, cashew nuts, sugar cane, pineapples, peach palm and a variety of cultivated tree fruits are also grown and sold in the town of Mitú. They make excellent hammocks of palm (Mauritia) fibre and patterned baskets from the rind of a wild species of Ischnosiphon. Painted bark cloth is used to make ceremonial masks and is now often sold for cash. All Kubeos are avid hunters; the shot-gun has supplanted the blow-gun, but they still fish with poisonous plants, bow and arrows, hand nets, fish hooks and elaborate weirs; fishing is mainly a seasonal occupation. Shamans practise an ancestor cult; especially noteworthy is the Yuruparí Dance when ancestral spirits are evoked by special sacred bark trumpets which women may not see on pain of death. They believe that the human spirit leaves the body during sleep and wanders nearby. Death occurs when a shaman ''catches the spirit''; at death, the

25

spirit leaves the body permanently but, unless driven away by certain rituals, may hover nearby. The shaman, as in other Tukanoan tribes, is believed to become a jaguar at will, especially when he takes caapi and can then diagnose and cure sickness, control weather, locate lost or stolen articles, affect fertility, even cause eclipses. There are several stages of medicine-men -- from the ordinary to the all-powerful who possess unlimited supernatural powers.

Wild fruits, gathered for food, also play an important role in an ancestor cult led by the shamans. The men bring in baskets of fruit to the sound of sacred bark trumpets. The trumpets represent the spirits of the ancestors, and this Yurupari cult was once common to all the Tukanoan tribes in the area. Manioc or cassava, taro, maize, sugar cane and peach palm serve to make a variety of beers brewed in canoe-like troughs and drunk at dances when each community entertains its neighbours. Food and drink is served in gourds grown by the women. After cleaning out the pith, the gourds are painted with sap from a tree and inverted over manioc leaves mixed with urine. The sap is then transformed into a hard black glaze. The same sap, painted on the wooden stools typical of all Tukano tribes, turns them red when exposed to the air.

Ceremonial dances serve as occasions for visiting and socialising and as religious rituals. Each group has its own varieties of coca and caapi which the men cultivate in their gardens. The exchange of coca is an important social act and virtually every adult male chews coca. Caapi, the basis of magico-religious and curative ritual and ceremonies, is drunk for the visions it produces as the men dance and chant. The shamans prepare and serve the caapi and the visions are interpreted as a direct experience of the spirit world described in myth.

The shamans mix different varieties of caapi together and dip leaves with holes in them into the brew. By examining the suspended films of liquid, they are able to control the strength and composition of the caapi to produce different effects. Some brews are quick acting, some are long lasting, some produce visions tinged with blue and others are red and fiery.

The Rio Piraparaná is blocked by beautiful but dangerous falls and this, together with the dangerous reputation of its inhabitants, has meant that tribes like the Makuna, Barasana and Taiwano still live in their traditional houses and remain relatively undisturbed by outside influences. The malocas of the Makuna are round at the base with a pitched roof made from woven caraná palm leaves *(Mauritia Carana)* on a wooden framework. Big triangular smoke-holes in the eaves serve to emit the sounds of great signal drums slung in frames by the door. Barasana and Taiwano malocas are rectangular in plan and often have designs painted on the sheets of bark which form the front wall. In this area, the men still wear g-strings of cotton cloth, and for ceremonies they add a white bark-cloth apron painted with a red pigment derived from the leaves of a cultivated vine *(Bignonia Chica)*. The same pigment, along with red from the seeds of urucú *(Bixa Orellana)*, is used as face paint for beauty and protection from the sun. A blue-black skin dye is prepared from the leaves of a cultivated shrub *(Genipa americana)* and used at rituals to paint intricate designs on the arms and legs.

Men clear gardens in the forest, but the women grow and harvest most crops, except coca. Manioc tubers provide a versatile base from which a number of foods can be prepared in addition to flat cakes baked on circular pottery hot plates and eaten with fish or meat supplied by the men who do the hunting and fishing.

Despite differences in the extent of external influence, the eastern Tukanoan peoples all share basic features in common. The earliest reliable descriptions of their way of life come from the great nineteenth century naturalists, Spruce and Wallace, who visited them in the 1850's. These early reports were followed by Koch-Grünberg's studies made during the first decade of the present century. Since then, Irving Goldman studied the Kubeo in the 1940's and further studies, by Kaj Arhem (Makuna), Christine and Stephen Hugh-Jones (Barasana), Jean Jackson (Bora) and Gerardo Reichel-Dolmatoff (Desana), were carried out from the mid-1960's onward.

The Siona are one of the western Tukanoan groups and live in the Comisaría del Putumayo in the region of Mocoa. They were first contacted by missionaries in 1632 and, like their neighbours the Kofán, they too have been heavily disrupted by the newly established oil industry. Despite these pressures, they still maintain many traditional customs, such as the use of the *cusma* or cotton cloak and the wearing of large quantities of glass beads. Though nose pins are no longer used, many still pierce their ears. They take great care of the hair, especially with certain oil-rich plants, and arm and leg bands are still worn, as is facial and body paint. Coca, caapi and yoco *(Paullinia Yoco)* are plants that play very important roles in their daily life. The western Tukanoans are culturally separated from the eastern groups and apparently have been isolated for hundreds of years. The ethnology of these Siona has recently been studied by Jean Langdon.

The Karib Indians in Colombia are known by the name Karijona. They originally lived in the upper Rio Caquetá and the headwater regions of the Rios Vaupés and Apaporis. Little of their history and life is known, except that they were semi-nomadic and war-like, feared by other tribes. In 1915, Whiffen estimated their population at 25,000. There are very few left, and these remnants now live in Miraflores on the upper Rio Vaupés and in the region of La Pedrera on the Rio Caquetá.

Even today, every visitor ... will be forced to acknowledge the Indian's subtle understanding of the ... flora and fauna if he will consider the significance of the names which the majority of animals, plants, rivers and mountains bear as their Indian inheritance. Like the faint echo of harpstrings, these names tell us of the lovable people who were once the rulers of the beautiful land and, at the same time, its children, born of its womb.

Konrad Guenther

Nature with folded hands seemed there,
Kneeling at her evening prayer.

Henry Wadsworth Longfellow

A large number of the trees forming these forests
are still unknown to science, and yet the Indians, these
practical botanists and zoologists, are well acquainted,
not only with their external appearance, but also with their
various properties ... it would greatly contribute to the progress
of science if a systematic record were made of all information
thus scattered throughout the land; an encyclopedia of the woods,
as it were, taken down from the tribes which inhabit them.

Prof. & Mrs. Louis Agassiz

Deus e grande, mas o mato e maior (God is great, but the forest is greater)
is an ancient saying among the Amazonian *caboclos*. May that forest be
greater than the encroaching evil of modern, destructive civilisation!

Harald Sioli

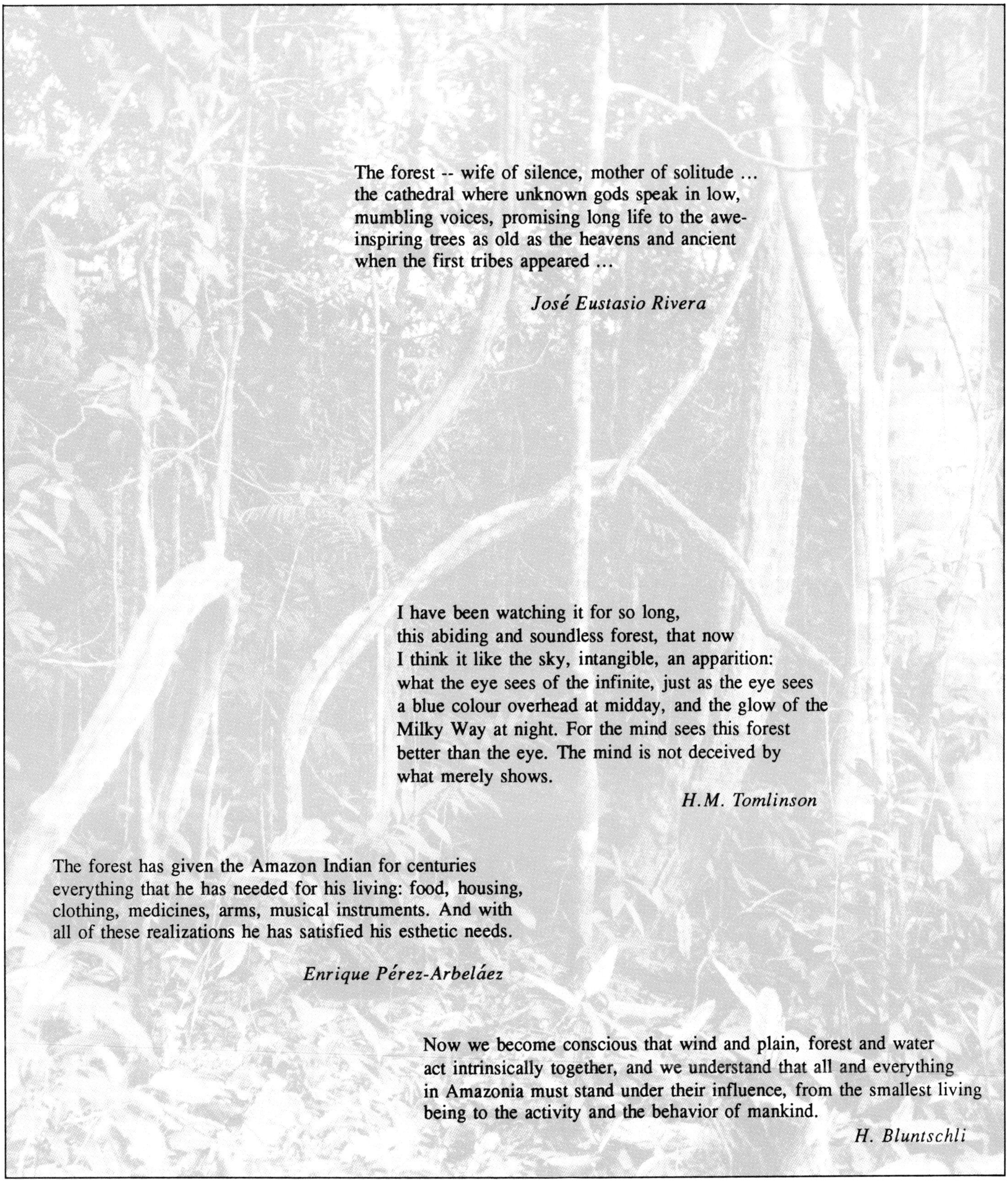

The forest -- wife of silence, mother of solitude ...
the cathedral where unknown gods speak in low,
mumbling voices, promising long life to the awe-
inspiring trees as old as the heavens and ancient
when the first tribes appeared ...

José Eustasio Rivera

I have been watching it for so long,
this abiding and soundless forest, that now
I think it like the sky, intangible, an apparition:
what the eye sees of the infinite, just as the eye sees
a blue colour overhead at midday, and the glow of the
Milky Way at night. For the mind sees this forest
better than the eye. The mind is not deceived by
what merely shows.

H.M. Tomlinson

The forest has given the Amazon Indian for centuries
everything that he has needed for his living: food, housing,
clothing, medicines, arms, musical instruments. And with
all of these realizations he has satisfied his esthetic needs.

Enrique Pérez-Arbeláez

Now we become conscious that wind and plain, forest and water
act intrinsically together, and we understand that all and everything
in Amazonia must stand under their influence, from the smallest living
being to the activity and the behavior of mankind.

H. Bluntschli

The Ajaju [is] a beautiful river, totally different from the others of the region, flowing in majestic curves at which appear enormous and fantastic cliffs standing out like ruins of feudal castles.

Phanor James Eder, *Colombia* (1913)

Sentinels in a lonely land
Cerros de Chiribiquete, Rio Ajaju, Vaupés

Years ago, the Ajaju -- in reality the source of the Rio Apaporis -- was populated by the Karijonas, a warring and cannibalistic Karib tribe. The terror of all other Indians within weeks of paddling distance, the Karijonas, forty years ago, fell into a bloody internecine war. This sent the few survivors, still further decimated and weakened by small-pox, to the headwaters of the Rio Vaupés, five long days to the northeast through the forest, and to the Rio Caquetá at La Pedrera, a thousand miles down the Apaporis. The Ajaju was left alone and silent, and all that reminds its glorious mountains of the proud race that once encamped at their bases are potsherds and charred faggots.

... like ruins of feudal castles

Here and there, sprouting up from the flatness like relic islands in a sea of crystalline emerald, are strange mountains clad with a thick vegetation.

Enrique Pérez-Arbeláez, *Vaupés* (1950)

Cerro de la Campana
Rio Ajaju, Vaupés

Remnants of a once continuous highland that stretched from the Guianas across southern Venezuela and nearly to the Andes in Colombia, the isolated quartzitic mountains of eastern Colombia are sentinels of a mysterious past.

The Cerro de La Campana (Bell Mountain) is one of the westernmost vestiges of these hills and is so strikingly awesome that it is wrapped in legend in the Indian mind. All Indians believe that fierce thunderstorms and torrents can be caused by beating upon a thinly eroded slab near the summit. When struck with another stone, it sends forth a bell-like tone. Caves near the base of La Campana have Indian paintings and scratching of animals on some of the walls, remains probably of the now nearly extinct but formerly numerous and warlike Karijonas who once dwelt in the basin of the upper Apaporis.

... relic islands in a sea of emerald ...

The most profound rest and quiet reigned at this height, which was only now and again interrupted by the rustling of the foliage set in motion by a breath of wind.

Richard Schomburgk, *Travels in British Guiana* (1922)

Ancient sandstone mountains
Rio Karurú, Vaupés

Nothing could be so peaceful as the summits of the ancient, isolated quartzitic mountains that are scattered throughout the Colombian Amazonia. Remnants of a once continuous mountain stretching from the Guianas, through Venezuela and into Colombia, they stand as monuments of by-gone ages and are older than the Andes and of the greater part of the Amazon basin. Here on the flat-topped heights the botanist is in literally a ''seventh heaven,'' for the flora of these mountains is strange, endemic and still mostly unstudied.

The most profound rest and quiet ...

... it is streaked with white, yellow and pink, perhaps from decomposition of its surface, but there seems also to be there an unusual proportion of mica in the granite. As very little water runs over this part it is not clad, as is most of the rest of the rock, with the same blackish Conferva which invests all the exposed granite in this region.

Alfred Russel Wallace (Ed.), Richard Spruce,
Notes of a Botanist on the Amazon and Andes
(1908)

Granitic mountain of the Brazilian shield
Rio Vaupés, Vaupés

Whereas the isolated mountains that are scattered throughout the western part of the Colombian Amazon are metamorphosed sandstone, those in the more eastern part near the frontier with Brazil are granitic and are resting on the Brazilian Shield. They are usually rounded with steep sides streaked with various colours. Their flora is limited in number of species and is not rich in endemics, differing in this respect from the sandstone mountains the vegetation of which is related closely to that of the *tepuís* of the Venezuela-Guianan land mass.

... streaked with white, yellow and pink ...

... there arose a blue mountain-top -- Mount Yupati, which forms the first great cataract of the Yapura and gives it its name.

> Theodor Koch-Grünberg, *Zwei Jahre unter den Indianern* (1910)

Cerro Kupatí or Cerro de La Pedrera
La Pedrera, Rio Caquetá, Amazonas

One of the most easily accessible sandstone mountains of Amazonian Colombia and a well known locality of high endemism, the Cerro de La Pedrera has not yet been thoroughly studied. Botanists have collected its flora only several times, each one of which has found plants known nowhere else. The earliest and classic collection was made by the famous German botanist von Martius in 1828. The mountain is so important to science that there have been suggestions that it should be protected as a biological reserve.

... there arose a blue mountain-top ...

The following morning was to greet us once more with one of those faery-like tropical landscapes, to which the eye of the Northerner finds expression in exclamations of surprise ... ahead of us ... there rose ... a sparsely wooded, isolated hill with innumerable white spots sparkling through its dark and refreshingly verdant carpet.

Richard Schomburgk, *Travels in British Guiana* (1922)

*Forests at the base of Cerro Isibukuri
Rio Kananarí, Vaupés*

Cerro Isibukuri rises abruptly from the forest floor, presenting to the western sun glistening white and red quartzite cliffs. Numerous are the ribbon-like falls that plunge from the summit to feed the coffee-brown Kananarí with their cold waters. No one has as yet satisfactorily explained the curious hue of so many of these streams arising from sandstone massifs. Why such waters are clear, without suspended silt and mud, can easily be understood. Does the brownish dye have its origins in tannins and other substances leached out from the drifts of leathery leaves which strew the sandy floor of the light caatinga forests surrounding these quartzite massifs? Whatever its origin, it is certain that the water acquires its strange colour while it stands in pools and swales in these light, grove-like forests.

... faery-like tropical landscapes ...

A marvelous sight, quite strange to one who has seen only thick jungle for so many months! The whole surface is beset with rocky ledges. A sparse vegetation such as I have never seen ekes out a wretched existence in the cracks of the rocks. Low, stunted bushes of a sickly growth, candelabra-like flowering trees with leathery leaves bunched at the tips and here and there bearing solitary many-coloured flowers. Far to the south one can catch a glimpse of the mountains of the upper Caiary [Vaupés] from which the legendary Taku, the dwelling place of demons, stands out in sharp relief with its rugged cliffs.

Theodor Koch-Grünberg, *Zwei Jahre unter den Indianern* (1910)

The Savannah of Yapobodá from the air
Headwaters of the Rio Kuduyarí, Vaupés

The Colombian Amazon region has innumerable ancient quartzitic mountains, most of them flat but some grotesquely eroded. They are the westernmost remnants of the once continuous Venezuela-Guiana land mass. With highly endemic floras related to those of the Roraima-like mountains in southern Venezuela, they present to the botanist intensely interesting plants specialized for living on almost soil-less sandstone and enduring conditions of severe drought, notwithstanding the usually heavy rainfall. To the Indian, they represent areas of mystery and are the subject of many native superstitions and legends; they are also the sources of many plants with extraordinary uses for which the Indians will gladly travel long distances to collect material for medicines, poisons and other purposes.

A marvellous sight …

In reality, the Indians are right! They are "rock-houses," but not made by the weak hand of man. All-powerful nature was the master-builder. How many thousands of years must have passed over them since the floods washed out these caves and shaped these gigantic labyrinths. For clearly they lead off with more entrances under the whole savannah. How many thousands of years must have been needed to finish such mighty works.

> Theodor Koch-Grünberg, *Zwei Jahre unter den Indianern* (1910)

Sandstone caves
Savannah of Yapoboda, Rio Kuduyarí, Vaupés

The discovery of large tunnels and caves in the ancient sandstone mountains and savannahs of the Vaupés is always the source of grave wonder amongst the Indians of the region. I have never been able to learn of any origin-legends, but I do not doubt that the older folk in some of the tribes look upon them with awe and have beliefs or stories which try to explain their origin and meaning. Some of these caves have drawings of animals -- both recognisable and imaginary ones -- engraven or painted on the walls.

... not made by the weak hand of man.

The language is sometimes called Camsá; the principal tribe is the Sibundoy ... there are said to be some 1,700 Sibundoy ...

> J. Alden Mason: "The Languages of South American Indians" in J.H. Steward (Ed.), *Handbook of South American Indians* (1950)

The Sibundoy Valley
Putumayo

This mountain-encircled valley, some 8,500 feet in the Andes of southern Colombia, is situated near the source of several of the Colombian tributaries of the Amazon. It is a cool, fertile region inhabited by two Indian groups -- the Sibundoy and the Inga: the former speak a language that has not yet been satisfactorily classified, but which may belong to the Chibchan family; the latter speak a Ketchwa dialect related to the language of the Andean Indians of adjacent Ecuador.

The Valley of Sibundoy is noteworthy for the high endemism of its flora. The Kamsá medicine men are incredibly knowledgeable concerning the properties of plants, and they employ a large number in their pharmacopeia. They also use many hallucinogenic plants in their diagnosis and treatment of disease. While these Indians are now very acculturated from the presence of missionary contact for nearly a century, they still conserve their language and many traditional elements of their culture.

... some 1700 Sibundoy ...

The largest river in the world runs through the largest forest ... a forest which is practically unlimited -- near three millions of square miles clad with trees and little else but trees ...

Alfred Russel Wallace (Ed.), Richard Spruce,
Notes of a Botanist on the Amazon and Andes
(1908)

The Amazon River above Leticia
Amazonas

Some 2,000 miles from its mouth, the Amazon still is a mighty river winding and twisting through interminable forests and making and then washing away islands and even parts of the bank and backland. It is along this type of river -- rich in fish and teeming with other wildlife -- that the heaviest populations dwell.

The largest river in the world runs through the largest forest.

... the river is in a predatory mood. The banks are attacked and pieces torn away ... and are swept away in the form of greyish yellow soup .... Whole patches of forest may be clawed away with huts, animals and men still on them .... The dense green walls of the forest are ripped down when a bank breaks. Trees droop into the water .... Like a giant's naked sinews the lianas are festooned across the eerie gaps .... The solidity of the land itself is only appearance.

Emil Schulthess, *The Amazon* (1962)

Forest and river near Leticia
Rio Amazonas

Upstream from Leticia, where the Amazon is very narrow, the river once again widens out and flows through mazes of islands. At this point, some two thousand miles from its mouth, the river often measures several miles in width. Its depth varies greatly between the highest level of the rainy season and the lowest of the dry. In 1945, I measured the rise and fall of the river at Leticia and found the variation to be 46 feet. In the rainy season, all of these low, sandy islands and much of the land, back for great distances from the banks, are under many feet of water. It is true that the Amazon is then a huge lake, not a river. This annual fluctuation has left a most definite mark upon the people of the area and has given rise to what has most appropriately been called an amphibious race of man.

The Amazon River itself and the Putumayo are the only so-called "white water rivers" in the Colombian Amazon -- rivers carrying heavy loads of silt that colour the waters a yellowish brown, quite in contrast to those known as "black water rivers" that carry little or no silt. The banks of the Amazon and Putumayo are generally low and flood annually, leaving silt which refreshes the soil. On these floodable banks and the numerous low islands the dominant tree is guarumo, several species of Cecropia. Most of the settlers who come to the Amazonian regions from the interior of the country choose to live and practice agriculture along these banks, partly because the deposition of silt each year provides the land with new nutrients brought down from the Andean highlands.

... the river is in a predatory mood.

A few miles before you reach the great fall ... large balls of froth come floating past you. The river appears beautifully marked with streaks of foam, and on your nearer approach the stream is whitened all over.

> Charles Waterton, *Wanderings in South America* (1825)

Early morning silhouette
Rio Apaporis, Vaupés

The daytime temperature in the Colombian Amazonia never reaches unbearable degrees, and the nights are delightfully cool in contrast. In the very early hours of the morning, the thermometer may hover in the upper sixties so that a remarkable blanket of condensation gently covers the landscape. As the sun tries to break through the thick curtain of haze, the sense of unearthliness is often striking. This is heightened when, as always happens just below the many falls and rapids, balls of froth churned up in the deafening tumult of the usually placid waters float silently down -- the only movement in an otherwise ghostly scene.

.. large balls of froth come floating past ...

The water was still and clear as glass: the trunks of the
trees stood up from it, their branches dipped into it; and
as we wound in and out among them, putting aside a
bough here and there, or stooping to float under a green
arbor, the reflection of every leaf was so perfect that
wood and water seemed to melt into each other, and it
was difficult to say where one began and the other ended.
Silence and shade so profound brooded over the whole
scene that the mere ripple of our paddles seemed a
disturbance.

Professor and Mrs. Louise Agassiz, *A
Journey in Brazil* (1868)

*Denseness of a forest rill
Rio Kotué, Amazonas*

The masses of epiphytes -- plants growing upon other plants but from
which they derive no sustenance -- in the Amazonian vegetation never
fail to awe the visitor. Many are the groups of plants that run to
epiphytism, but the orchid and the pineapple families (Orchidaceae
and Bromeliaceae) are, together with the aroids (Araceae), perhaps the
most conspicuous and numerous in individuals and species. Some of
the bromeliaceous species hold water in their leaf bases, and whole
faunas of insects, spiders and minute frogs find in this curious habitat
their homes. The impression of density is heightened by this striving
for a place in the sun, and many are the true hanging gardens which
epiphytism creates for the greater efficiency of nature.

... wood and water seemed to melt into each other ...

Wherever the eye turns, it meets with fresh surprises. Here a mighty current rushes in between the rocky cliffs to disappear as if by magic into an unnoticed gulf; there a huge mass of water is ever on the whirl in a funnel-shaped cauldron formed of giant boulders.

Richard Schomburgk, *Travels in British Guiana* (1922)

The tunnel of Jirijirimo
Rio Apaporis, Vaupés

The mighty Apaporis, after it tumbles over the Falls of Jirijirimo, enters a long and narrow chasm walled in by high, vertical cliffs. At one point, the whole river disappears into a tunnel, flowing tranquilly and deep through the curious fault. This is a place of awful mystery to the Indians of the area who, except for the medicine-men, never travel through the chasm, and the tunnel is known to them only through hearsay.

Wherever the eye turns ... fresh surprises ...

... any extraordinary or inaccessible rock is always said
to be inhabited by monstrous animals.

> Everard F. Imthurn, *Among the Indians of Guiana* (1883)

Milestone in the Rio Guainía
Raudal del Sapo, Rio Guainía, Vaupés

There are many rocks at the rapids in the numerous rivers of the
northwest Amazon that serve as ''milestones'' for the natives. The
photograph illustrates the rock that looks like a sitting frog and which
gives the name to the adjacent rapids: Raudal del Sapo (''Frog
Rapids'') in the Rio Guainía. These curiously formed rocks serve not
only as points of reference to the Indians, but each one has a myth or
story usually explaining its supernatural origin.

... inhabited by monstrous animals

At our feet, there thundered and foamed the raging body
of water wrestling with itself, and the rocky cliffs at the
same time that, like a crater belching forth the stony
fragments freed from its entrails, it spouted and scattered
the spray clouds and foam-flakes into the air to build a
continual change of innumerable rainbows only to
disappear as rapidly as they were formed.

> Richard Schomburgk, *Travels in British
> Guiana* (1922)

The Falls of Yuruparí (the Devil)
Rio Vaupés, Vaupés

These beautiful waterfalls are the farthest upstream in the Rio Vaupés
which, below this point, is strewn with rapids and waterfalls. The
river above these falls is placid, wide, with a gentle flow. In the upper
regions of the river many white settlers have come in to live, whereas
below this falls the population is primarily Indian. The British
naturalist, Wallace, went upstream as far as this magnificent falls in
the middle of the last century. Because of the geological substratum
which changes at Yuruparí, the flora has a noticeably different
composition than that of the lower regions of the river.

60

... the raging body of water wrestling with itself ...

The roaring of the water was dreadful; it foamed and dashed over the rocks with a tremendous spray ... threatening destruction to whatever approached it. You would have thought, by the confusion it caused in the river, and the whirlpools it made, that Scylla and Charybdis, and their whole progeny, had left the Mediterranean and come and settled here.

Charles Waterton, *Wanderings in South America* (1825)

Falls of Yayacopí
Rio Apaporis, Vaupés

The thundering falls of Yayacopi strike awe into the hearts of the Indians of the region, accustomed as they are to the titanic forces of angry waters everywhere in the Apaporis basin. This awe manifests itself in a legendary belief amongst the Makunas who dwell downstream from the Yayacopí that an early witch-doctor of the tribe, determined to protect his people from a fierce and ever-warring tribe that once inhabited the headwaters of the Apaporis and whose frequent depredations almost caused the extinction of the Makunas, took such a draught of the narcotic yajé that he was able to commune with the spirits friendly to the Makunas for a whole week. During this week, a scheme was elaborated, whereby the gods raised up mighty hills which stretched across the Apaporis to form the series of three treacherous and impassable rapids, the greatest and lowermost of which is Yayacopí. These formidable barriers and certain powerful hexes which were laid over the whole area prevented the up-river enemies from continuing their raids.

Though Makunas frequently paddle several days to fish in the richly stocked whirlpools at the foot of Yayacopí, it is true that they never willingly journey upstream farther than Yayacopí, even though the fierce tribes of the headwaters (the reputedly cannibalistic Karijonas) have, through internecine strife and disease, been reduced to a mere handful and have left the Apaporis entirely. And Makuna witch-doctors frequently make pilgrimages to this great waterfall where they practice elaborate incantations.

The roaring of the water was dreadful ...

At last, we came in sight of the great cataract itself, and it was one of the most impressive objects I have ever seen ... The noise of the howling waves was so great that we were almost deafened. My Indians, who were always very susceptible to natural phenomena, seemed to regard the great cataract as a revelation of divinity.

William Montgomery McGovern, *Jungle Paths and Inca Ruins* (1927)

The great Falls of Jirijirimo
Rio Apaporis, Vaupés

Below the mouth of the Kananarí, the great and unhurried Apaporis narrows its half-mile width to one hundred feet, struggling over fretful reefs to a seventy foot fall and nine mile chasm. Jirijirimo is a sacred place for the Indians of the Kananarí, a show of unbridled might, awesome as "a revelation of divinity".

... a revelation of divinity.

The rising sun was covered with that white transparent cloudy fleece so peculiar to the rainy season, and now and then cast its rays upon the millions of rain drops glittering like diamonds hanging from the trees, and which in association lent an enchanting freshness to the whole of the vegetation.

Richard Schomburgk, *Travels in British Guiana* (1922)

Morning mist over the Falls of Jirijirimo
Rio Apaporis, Vaupés

Early morning, especially in the rainy season, is a beautiful time of day. One feels that all is well with oneself and the world. The first rays of light call forth gentle jungle voices that speak only in the dawn. There is a feeling of living within a fleecy dome, a welkin of unreality which gradually fades away with the strengthening sun.

Looking downstream on the Rio Apaporis at early dawn just above the Falls of Jirijirimo, one experiences these feelings; and the sensation of other-worldliness is heightened by the constant dull thunder of the water throwing itself over the tremendous drop. The mushrooming mist rises from the falls and hangs ominously as if to cover the wild beauty of the churning river from the sun's inspection.

Can we wonder that the Indian holds this place in such awesome wonder?

... white transparent cloudy fleece ...

Beneath me were the rapids tumbling ... with a noise which we had heard an hour before reaching them. Then spread out the glorious river, empurpled with the rays of the departed sun ... the waters everywhere circling and eddying or running rapidly over some sunken ledge of rock.

Alfred Russell Wallace (Ed.), Richard Spruce,
Notes of a Botanist on the Amazon and Andes
(1908)

The beginning of Jirijirimo Falls
Rio Apaporis, Vaupés

The great Falls of Jirijirimo must be considered one of the natural wonders of Colombia. It begins its tumbles over several steps formed by the rocks before it gets to the vertical fall. The turbulent frenzy of the waters as they rush over these steps as if in a hurry to reach the fall itself is in great contrast to the leisurely flow of the wide river itself. In view of the natural beauty and complexity of this waterfall there can be little wonder why the Indians ascribe a supernatural origin to it.

Beneath me were the rapids tumbling …

It is through this mountain that the Caquetá is obliged to force its way. These white cliffs, made up of rock which is riven both vertically and horizontally, look like ramparts thrown up by giants.

J. Crévaux, *Voyages dans l'Amérique du Sud*
(1883)

Mountains and cataracts of Araracuara,
Rio Caquetá, Amazonas

Araracuara is famous and infamous. One of nature's most awesome sights, its terrible rapids and rushing chasms have claimed lives in the hundreds. Utterly impassable by river, Araracuara marks the westernmost point of penetration of the renowned German plant-explorer Carl Friedrich von Martius in 1820. Von Martius discovered many rare and weird plants on the sandstone hills at this locality: a large number of them are known only from Araracuara, and some have not yet been collected again by botanists. Araracuara is still one of the richest treasure-houses that beckon to the naturalist. And downstream from Araracuara there empties the great and unknown Yarí River. I have flown over the Yarí basin many times and know that untold surprises in new plants await the botanist and that unknown groups of Indians, living in curious conical houses, await study.

For many years, the Colombian government maintained a noxious penal colony at Araracuara where the nation's worst criminals were sent, isolated from the world by impenetrable forests and invincible rapids making for a more secure confinement than locks and bars. The Indians fled from this region. Now the penal colony no longer exists, and its place has been taken by a model agricultural experiment station; the Witotos and Andokes have happily returned.

... ramparts thrown up by giants ...

The Barasana have always lived here; they were born
here. They and the Taiwano have always been friends.
They both have the same god who has power over all
things. It is he who made Nyi in his own likeness; the
drawings have been here forever, and this is the only
place where they occur.

> [A Barasana Indian quoted by] Brian Moser
> and Donald Tayler, *The Cocaine Eaters*
> (1965)

The god of the river or of water, Nyi
Río Piraparaná, Vaupés

This rock-engraving represents probably the most elaborate one in the
northwest Amazon. It is carved in extremely hard granite and in places
the etching is half an inch deep. The total height of the carving is five
and a half feet. According to Tukano mythology, this petroglyph is
associated with the Sun Father's visit to the House of Waters when he
gave these Indians the hallucinogenic plant caapi. It is situated on the
equator where mankind had its origin. This engraving, connected with
the origin myth of the Tukanoan peoples, claims a very special
reverence on the part of the Indians of the Río Piraparaná.

... the drawings have been here forever ...

Hardly another relic of past ages in South America has called forth amongst scholars so many varied and contradictory explanations as have the drawings and figures which one finds graven on the rocks by human hand.

Theodor Koch-Grünberg, *Sudamerikanische Felszeichnungen* (1907)

Rock engravings in granite
Rio Piraparaná, Vaupés

It is true that the origin and significance of most of the rock engravings in the rivers of eastern Colombia are lost in the darkness of past ages. We do know, nonetheless, that some of the more elaborate ones represent spirits or gods of fertility.

In the Vaupés, the Rio Piraparaná seems, more than any other region, to abound in rock carvings which, to the Indian of the area and to traveller alike, are obviously connected with phallism. The Indian will always say, as do the Tatuyos of the Piraparaná of this enormous and complex petroglyph, that the gods of ages ago drew them and endowed them with supernatural powers. Some natives will tell the traveller that these are indeed connected with human fertility and have as their dwelling place the rivers, since it is the river waters that make the region give birth to fruits every year.

... so many varied and contradictory explanations ...

Many of the relatively inaccessible human figures placed
along streams and by waterfalls ... are probably also
symbolic of the spirits believed to reside in those places.
Even today, it is difficult to persuade some natives to
approach these drawings.

Irving Rouse, ''Petroglyphs'' in *Handbook of
South American Indians* (1949)

*Inaccessible rock-engravings
Rio Piraparaná, Vaupés*

It is impossible to say what Indians of ancient times created these
rock-engravings nor how long ago they were made. Most of them, at
least those along the Rio Piraparaná, are located in well-nigh
inaccessible places and were engraved on hard granite rocks usually
at swiftly flowing narrows in the river or at rapids or waterfalls.
Whoever these ancient artisans were, they obviously had no metal
tools. The whole mystery of their creation is difficult to understand.
It is small wonder that today's inhabitants assign supernatural origins
to these works of art.

... symbolic of the spirits ...

Upon inquiring from the natives as to who had made them we everywhere received the reply: "Our forefathers, when the immense waters still covered the earth and they navigated the mountains in their corials".

Richard Schomburgk, *Travels in British Guiana* (1922)

Rock engravings of fertility symbols
La Pedrera, Rio Caquetá Amazonas

Uncovered only during the dry season, the slanting, smooth, brown sandstone slabs in the Caquetá River at La Pedrera are strewn with engravings obviously connected with ancient fertility rites. Even amongst the Miraña and Yukuna Indians of the region today exists a belief that these were carved by virgins whose curious duties devolved around the increase of fertility of ancient tribes once inhabiting the region. Few Indian men -- save the witch-doctors -- have ever seen these carvings. If they must transit this part of the river in their canoes, they choose the opposite bank where there are no rocks. For every Indian man will assure the traveller that so powerful are these engravings still that he who looks upon them will produce only female offspring, nevermore begetting a male child. Most of the engravings at La Pedrera seen by the botanist-explorer Von Martius in 1828 are fast disintegrating and washing away.

… when the immense waters still covered the earth

The decorative patterns ... are almost wholly derived
from drug-induced inner light experiences.

> Gerardo Reichel-Dolmatoff, *Beyond the Milky
> Way: Hallucinatory Imagery of the Tukano
> Indians* (1978)

Engravings in metamorphosed sandstone
La Pedrera, Rio Caquetá Amazonas

It is probably extremely significant that the designs in many of the
aspects of modern Indian artistry in the northwest Amazon are similar
to or the same as those found in many of the rock-engravings. Studies
have indicated that these designs -- on painted houses, on ceramics and
in other household applications -- are suggested by visions
experienced during the intoxication produced by caapi *(Banisteriopsis
Caapi)*, one of the important sacred hallucinogenic plants of the
region. There is no reason to doubt that the ancient artisans who made
these rock-engravings had used the same drugs and had the same
experiences as the natives of today.

... from drug-induced inner light ...

I have lately been calculating the number of species that yet remain to be discovered in the great Amazonian forests, from the cataracts of the Orinoco to the mountains of Matto Grosso ... there should still remain some 50,000 or even 80,000 species undiscovered.

Alfred Russel Wallace (Ed.), Richard Spruce,
Notes of a Botanist of the Amazon and Andes
(1908)

*Amazonian forest near the Falls of Jirijirimo
Rio Apaporis, Vaupés*

It is, of course, still too soon to give a definitive calculation of the number of species of plants that the great Amazon Valley harbours. Botanical exploration is still in its relative infancy. Those botanists who have worked in this incredible wealth of vegetation are inclined to believe with one of the greatest and most perspicacious plant scientists who have worked in the Amazon that perhaps the flora would reach some 80,000 species. If we realize that studies of the cryptogamic flora of this immense region has hardly begun, some vision of the incomparable wealth of the region in its plants may perhaps be realized.

In some parts of the Amazon Valley, species are becoming extinct as a result of uncontrolled human activity. What is unfortunately less widely recognized is that the Indians' knowledge of the properties of thousands of species -- acquired over millennia -- is disappearing faster than many plants as a result of advancing westernization of aboriginal peoples brought about by missionary activity, commercial penetration, road-building, tourism -- all leading to acculturation and the disappearance, often in one generation, of native lore, and in some areas of extinction of cultures or even of whole tribes.

... there should still remain some ... 80,000 species undiscovered.

All the pictures my imagination had painted in anticipation of the impression of a virgin forest would make on me sank like faded shadows into insignificance before the sublime Reality that discloses itself on entering it!

Richard Schomburgk, *Travels in British Guiana* (1922)

Undisturbed forest
Path from the Rio Karaparaná to the Rio Igaraparaná
Amazonas

When someone accustomed to the forests of the north-temperate zone first enters the welter of the estimated 80,000 species of plants in the Amazon forests, an immediate awe overtakes his thinking and his comprehension of the environment. Usually, the visitor from northern regions begins to regard the green cover with something best described as religious reverence. Then, his first practical impulse leads him to ask: "How can anyone think of destroying such a magnificent heritage that Nature has given us?" This is a question that should be asked of political and commercial personnel who have in their power the protection or destruction of the last humid, tropical rainforests of the world which mankind and nature must have to survive.

... the sublime Reality that discloses itself ...

The woods of this river are innumerable, so tall that they reach to the clouds, so thick that it causes astonishment ... Here ... is timber; here are cables made from the bark of a certain tree ..., here is an excellent pitch and tar; here is an oil ...; here they can make excellent oakum ... for caulking ships; and also there is nothing better for the string of an arquebuss; here is cotton for the sails; and here finally is a great multitude of people, so that there is nothing wanting ...

Father Cristoval de Acuña, *A New Discovery of the Great River of the Amazon* (1641)

A Makuna tree-climber working for a botanist
Rio Piraparaná, Vaupés

The problem of a naturalist in this dense and undisturbed tropical forest lies in his hesitation at felling a mature, majestic tree for a few specimens to be studied in our herbaria. One way around this problem is to engage the services of agile, young Indians who, with unbelievable ease, can climb the most inaccessible trees to cut off a single branch. These aboriginal helpers and their agility greatly reduce the botanists' need to fell a tree in its full vigour merely for an identification of its botanical identity. We must, therefore, give these Indians credit for their contribution to conservationism. Many of the younger Indians understand the reluctance to felling a mature tree and are willing to engage in the physical effort to climb and cut off a branch.

... the woods are so tall that they reach to the clouds ...

As the home of the red man harmonises with the forest, so does his canoe with the dark waters of the creek. He uses no ornament on his frail craft, so that whether it lies on the surface of the water, is hidden in a tangle of bush-ropes, or drawn up on the bank, it is hardly distinguishable from its surroundings ... In the gloom of the forest arcade, the canoe glides along as noiselessly as if it were a brown aquatic creature with a man's body and shoulders above water slowly guiding itself with one great flapper.

James Rodway, *In the Guiana Forests* (1911)

Barasana paddler
Rio Piraparaná, Vaupés

The interminably intertwined waterways of the hinterlands of the Vaupés are constantly traversed in small dugout canoes by the Indians who inhabit the region so sparsely. One of the most wondrous sights which can absorb the interest of a traveller is that of a naked Indian prowling silently in his dugout through the leaves of a flooded forest. The traveller who has not seen such a sight has not been fully initiated into life in the Amazon jungle.

... hardly distinguishable in the gloom of the forest arcade ...

The Amazon has attracted scientific investigators with an irresistible force ... [But] science in the Amazon has not yet become systematized. It is barely now experiencing orderly, penetrating and exhaustive research and publication.

Enrique Peréz-Arbeláez, *Hilea Amazonica Colombiana* (1949)

The "devil's garden"
Puerto Limón, Rio Caquetá, Putumayo

There are in the western Amazon forests isolated areas where almost nothing but *Selaginella* will grow, with the exception of small trees or bushes of the Coffee Family which are botanically known as species of *Duroia.* These rather open spots are called "devil's gardens" by the natives. They remain a scientific mystery. It has not been determined what causes the inability of the surrounding forest to thrive where *Duroia* is abundant. *Duroia* is usually inhabited by ferocious ants, the explanation of some of the Indians. Other believe that the *Duroia* exudes or exhales something toxic to most other vegetation. Whatever the cause, there is here an interesting and unsolved ecological enigma.

Amongst numerous tribes in the Colombian Amazon, there is a curious custom of using the fresh, pliable bark of *Duroia,* bound tightly around the biceps, to induce a cosmetically esteemed ornament. The bark is caustic and raises blisters. Several days following the blistering a blue-black ring in the skin appears; this ring persists for a month or longer.

Science in the Amazon has not yet become systematized.

When I turned to look up at the trunks and branches, there smiled at me through the semi-obscurity prevailing over the whole forest, fresh, tumid mosses and lichens, pretty ferns, most beautiful orchids and aroids, the white or greenish aerial roots of which almost reached to earth.

Richard Schomburgk, *Travels in British Guiana* (1922)

Aerial roots of an aroid
El Encanto, Rio Karaparaná, Amazonas

One of the characteristics of the complex Amazon forest is the great number of epiphytes, plants that grow upon trees and shrubs seeking more light from the sun. They do not harm the plants upon which they grow for, unlike true parasites, their tissue does not enter the "host." There are many groups that specialise in this kind of epiphytism: notably the orchids, members of the Panama straw-hat and pineapple families, the ferns and the aroids, not to mention the many mosses and hepatics, algae and lichens. The aroids especially send down from the tops of the tallest trees great bands of aerial roots to absorb moisture and nutrients from the soil. The Indian has found that these roots, extremely strong and flexible, are useful for binding together the rafters of their houses.

... the white or greenish aerial roots ...

Indistinguishable tree from creeper, parasite from supporter, all are clamoring for space and light and air.

Millicent Todd, *Peru, a Land of Contrasts*
(1918)

An inundated forest
Río Cahuinarí, Amazonas

↗ curare

The easily and heavily flooded forests are the densest and most difficult of access of the usually pleasant and inviting areas of vegetation offered by the northwest Amazon. Yet in this tangled growth there live some of the most useful of the medicinal and toxic plants utilized by the Indian societies and which, when fully investigated, may provide more advanced pharmacopoeas with valuable chemical compounds for treating diseases that still cannot be cured by tools of modern medicine.

... clamoring for space ...

Vegetation invades everything. It shoots out over the water, covering it with lovely forms. Hardly a growing thing can get its impulse directly from the soil. That was long ago preempted. There must be other things to grow upon or in.

Millicent Todd, *Peru, a Land of Contrasts*
(1918)

A watery glade in the Colombian Amazon
Upper Rio Caquetá, Amazonas

The overwhelming luxuriance of the forests of the northwest Amazon fools the casual traveller into the belief that the soils must be amongst the richest of the world. Nothing could be farther from the truth. The vegetation is lush, but it is a vegetation which through the ages has become adapted to poor soils and conditions extremely unfavourable to the normal development of most plant life. One of the gravest errors -- and one which is constantly trapping the unwary writer -- is the conviction that the forests of the Amazon are a great waste and should be cleared away for the immediate settlement of immigrants from overcrowded parts of the earth's surface. The Amazon -- especially the northwestern portion -- can never support large populations of civilised man.

Vegetation invades everything.

In the Rio Negro, the river which waters Manáos, salt is extracted in a great quantity from certain plants which grow upon rocks standing in the midst of the strongest of fresh water currents.

Baron de Santa-Anna Nery, *The Land of the Amazons* (1901)

River weeds covering rocks
Rio Apaporis, Amazonas

The Falls of Jirijirimo begin with short cascades where a half-mile-wide stream narrows into a sixty-foot chasm through which the water churns over storeyed rocks and forms seething cauldrons where only the hardiest of plants, like the podostemonaceous species here shown, can get a foothold and survive. Even in such inhospitable of spots, Nature takes over with an abundance of plant growth.

The podostemonaceous plants have tough, alga-like leaves that come out at the height of the rainy season and clothe the rocks where the flood will reach its fullest. The tiny white flowers have blossomed in time to set ripe fruit for the fullest sweep of the waters. The natives along the Apaporis gather the leaves which they reduce to ashes to make a kind of salt employed as we use table salt. Our own sodium chloride, greatly sought from travellers by the Indians, does not exist in the Amazon, and the potassium-rich ashes of the river weeds have for centuries filled the need which we satisfy with salt. Indeed, the Makunas call our salt by the same word -- *moo-a'*-- which they apply to these weeds of the rapids.

... salt from certain plants ...

This palm is not only one of the most beautiful, but one of the loftiest in the country; the leaves are fan-shaped and form a large, round ball at the top of the stem ... It produces a great number of nuts, about the size of a small egg, covered with rhomboidal scales, arranged in a spiral manner; between these scales and the albuminous substance of the nut, there exists an oily pulp of a reddish colour, which the inhabitants ... boil with sugar and make into a sweetmeat ... they prepare from this pulp an emulsion, which when sweetened with sugar forms a very palatable beverage, but if much used is said to tinge the skin a yellowish colour.

> Berthold Seemann, *Popular History of the Palms and Their Allies* (1856)

Fruits of the moriche palm
La Chorrera, Rio Igaraparaná, Amazonas

The moriche palm *(Mauritia flexuosa),* a stately and tall tree growing in great abundance and often in almost pure stands in boggy areas, is one of the most useful members of the Amazonian flora. It supplies building materials for houses, carbohydrate food, cork for bottles, fibre for weaving, a fermented beverage and diverse other products important in tribal economies. There are numerous legends concerning this palm in the various tribes, and it enters into sundry beliefs and mythological tales.

Chicha de moriche, the alcoholic drink prepared from the yellow or orange, oil-rich flesh of the brown, egg-shaped, scaly fruit, is a valuable source of vitamin B1 in the diets of the northwest Amazon.

This palm has two names in the Colombian Amazon: *moriche* in the Vaupés and *cananguche* in the Putumayo and Amazonas.

... nuts covered with rhomboidal scales ...

From the number of spines with which these palms are clad, they have rather a repulsive aspect, yet nevertheless, and notwithstanding their being "armed to the teeth," it has not prevented man from approaching them and finding out about their various useful properties.

Berthold Seemann, *Popular History of the Palms and Their Allies* (1856)

*Spiny trunk of the cumare palm
Rio Popeyacá, Amazonas*

The *cumare* palm tree *(Astrocaryum vulgare)* -- known also as *chambira* in some parts of eastern Colombia -- is one of the forest's greatest gifts to the Indian. For its leaflets yield a wonderfully strong fibre from which the natives weave their netted hammocks and which has many other uses in their economy. Growing on the high, well drained sandy hinterland which the annual flood of the rivers cannot reach, the cumare tree is often found in dense pockets, and it is understandable that Indian houses are rarely built far from a good colony of this most useful of the palms.

... armed to the teeth ...

The women twist the thread with great dexterity. They sit on the ground, and taking two threads, which consist of a number of minute fibres, between the finger and thumb of the left hand, they lay them, separated a little, on the right thigh. A roll of them down the thigh, under the right hand, twists each thread; then, with a scarcely perceptible motion of the hand, the worker brings the two together, and a roll up the thigh makes the cord. A woman will twist fifty fathoms about the size of ordinary twine in a day.

William Lewis Herndon, *Explorations of the Valley of the Amazona* (1854)

Indian girls separating cumare-fibre
Rio Apaporis, Vaupés

Fibre is extracted from the leaflets of the cumare palm and, while still fresh and green, is rolled between the thigh and palm. The resulting fibre is unexcelled for hammocks and nets, and its preparation takes up many hours of a woman's life in this area.

The women twist the thread with great dexterity.

The palms, as a natural group, stand out among all other plants with remarkable distinctness and individuality. And yet this common character, uniting them so closely as a natural order, does not prevent the most striking difference between various kinds of palms. As a whole, no family of trees is more similar; generically and specifically, none is more varied, even though other families include a greater number of species.

Professor and Mrs. Lewis Agassiz, *A Journey in Brazil* (1868)

Feathery palms along a rapids
Rio Kananarí, Vaupés

In the Vaupés, the rivers are abundantly supplied with rapids and waterfalls. Alongside many of the rapids, the graceful *caranaí (Mauritiella cataractarum)* with its delicate leaves shades the rock-strewn banks. Because it is found only in this particular habitat, native belief holds that it was planted there when the gods, fighting each other, threw up the many rapids and falls as ramparts and that this happened before man ever came to the Vaupés from the Milky Way.

... distinctness and individuality ...

The *Victoria regia* is not an ornamental plant abundantly visible in the Amazon panorama. It lives in the seclusion of immense lakes where the water is calm and motionless. Its flower is white when young but gradually turns rose-coloured. Its name was given in homage to Queen Victoria of England.

Harald Schultz, *Isto e a Amazônia* (1964)

The aquatic wonder, Victoria amazonica
Leticia, Rio Amazonas, Amazonas

This largest of all water lilies lives in inlets, lakes and other bodies of still water in the Amazon and in British Guiana. Numerous large populations of this curious plant are preserved near the Colombian town of Leticia, where it is known by its Brazilian names *uapé* or *aguapé*. The local inhabitants employ the leaves to soften ulcers and infected wounds.

Although the correct botanical name because of priority of publication is *Victoria amazonica,* it is much better known by the more appropriate *V. regia,* named in honour of HRH Queen Victoria. One of the natural wonders of the Amazon, this lily is outstanding for its enormous leaves that may measure up to six and one-half feet in diameter and the huge, fragrant flowers that can reach a size of 18 inches across when fully open. The leaves, green above but reddish beneath, have an upturned margin five or six inches high and underneath are strengthened with a lattice-like network of thickened spiny ribs. The flowers, pollinated by a large beetle, are nocturnal, opening in the cool of the afternoon and closing by noon on the following day; in this short period, the 50 or more petals change from white to a dark pink.

Discovered in 1801, this beautiful plant was not well known for some 35 years. During the past century it has been avidly sought as a greenhouse ornamental by many botanical gardens and has earned the English name of Royal Water Lily.

... in the seclusion of great lakes ...

... that curious palm, the Piassaba, which produces the fibrous substance now used for making brooms and brushes ... it grows in moist places and is about twenty or thirty feet high, with the leaves large, pinnate, shining and very smooth and regular. The whole stem is covered with a thick coating of the fibres, hanging down like coarse hair, and growing from the bases of the leaves, which remain attached to the stem. Large parties of men, women and children go into the forests to cut this fibre ... I believe it to be a species of *Leopoldinia* ... and have called it *Leopoldinia Piassaba* from its native name in the greater part of the district which it inhabits.

Alfred Russel Wallace, *A Narrative of Travels on the Amazon and Rio Negro* (1889)

Bundles of piassaba fibre
Rio Guainía, Vaupés

One of the strangest products of the Amazon and Orinoco jungle is a coarse fibre which the Venezuelans call *chiquichiqui,* the Brazilians and Colombians *piassaba.* The fibres are the remains of decayed and fallen leaves. The piassaba palm grows only in the light savannah forest, typical of the white-sand areas of the upper Rio Negro area. When a tree is cleaned of its fibre, it cannot be profitably visited again for fifteen years. For many years, a lucrative commerce in piassaba has flourished in the Brazilian city of Manáos, the fibre being transported by river launch, around tremendous rapids and falls, the whole length of the Rio Negro. In recent years, piassaba has fetched such a high price in Bogotá that it has been commercially possible to fly the heavy conical bundles of the fibre from the Rio Negro to Bogotá, a four-hour flight. Piassaba gathering is a family industry, whole groups of Indians working through the four or five months of the dry season in the nomadic activity of clearing out the *piassabales* or piassaba forests in a given area which, the previous year, has been scouted. The Baniva and Kuripako Indians of the Rio Guainía are unusually adept at this work and piassaba grows almost invariably where there is no native rubber to provide a source of income.

... that curious palm, the *piassaba* ...

In the Indians' villages, about the houses, many hundreds of these trees may often be seen, adding to the beauty of the landscape and supplying an abundance of wholesome food ...

Berthold Seemann, *Popular History of the Palms and Their Allies* (1856)

Peach-palm guardians of an Indian site
Rio Kuduyarí, Vaupés

No one knows where the peach-palm *(Guilielma speciosa)* is native, but it has spread far and wide in tropical America. It has never been found truly wild in the Amazon. Clumps of this graceful tree, lifting its lacy fronds through the jungle, indicate sites of former Indian dwellings. Most of Colombia's Amazon tribes fence their houses with a circle of the *chontaduro* or *pupuña.* Its yellow or orange peach-like fruit, mealy and stocked with nutritious oils, provide one of the important foods of the area. It is eaten boiled or roasted; it is made into a fermented drink or *chicha;* it is prepared into a meal. So vital is this tree to the Indian that around its harvest he has built many of his ritual dances.

... the beauty of the landscape ...

Tropical forest villages range in population from 20 to
around 2,000 individuals, although 50 or 60 seems fairly
average.

Wendell C. Bennett in J. H. Steward (Ed.),
Handbook of South American Indians (1949)

Koreguaje settlement
Rio Orteguaza, Comisaría del Caquetá

Unlike most indigenous groups in the Colombian Amazonia, the
Koreguajes tend to live in villages or settlements of ten to fifteen
thatched houses very closely built to each other. It is partly because
of this closeness of tribal life that so many traditional beliefs, practices
and respect have survived in a part of the Colombian Amazon that has
been populated by western culture.

The Koreguajes still preserve traditional customs such as the
chewing of coca.

... villages range in population ...

The bark covering the front of many Indian houses ... gives the inhabitants, as we have seen on many occasions, a place to put their primitive art to use. Sometimes there are more or less characteristic figures of men, animals and objects of daily use or patterns used in weaving simply sketched in charcoal. Oftentimes, however, the walls of these houses are richly ornamented in many colours -- black, red, yellow, white -- with orderly designs.

> Theodor Koch-Grünberg, *Zwei Jahre unter den Indianern* (1910)

Kabuyarí house decoration
Rio Kananarí, Vaupés

Many are the Indian *malocas* or communal dwellings which, like this one amongst the Kabuyarís, proudly wear elaborate decorations painted on the bark of the outer walls. The decorations usually consist of human or geometrical figures copied from the rock-engravings found on boulders in the rivers and supposed to be harbingers either of luck or of misfortune. The paints most commonly employed are yellow, white, reddish or brownish and black. Made from earths and plants, they are surprisingly durable under the heavy rainfall of the Amazon skies.

Far from engaging in non-utilitarian decoration, the Indian is convinced that in thus adapting the rock-engravings to adorn the walls of his dwelling, he is protecting the house and its inhabitants from the wrath of evil spirits and forces operating constantly and ever needing propitiation.

116

... a place to put their primitive art to use ...

The natives ... think no more of destroying the noblest trees ... than we the vilest weeds; a single tree cut down makes no greater gap, and is no more missed than when one pulls up a stalk of groundsel or a poppy in an English cornfield.

Alfred Russel Wallace (Ed.), Richard Spruce,
Notes of a Botanist on the Amazon and Andes
(1908)

Tatuyo house
Rio Piraparaná, Vaupés

In general, the Indians in the basin of the Rio Piraparaná dwell far away from the main stream, at the headwaters of tiny creeks and brooks. This would seem to be a defense against possible molestation by the very infrequent white travellers, rubber workers or missionaries and partly because of mutual distrust amongst the sundry tribes inhabiting the region. The Indians of this river enjoy a reputation for fierceness and treachery -- a reputation certainly not deserved. They live in large communal houses isolated by days of travel through the forest from their nearest neighbours. One peculiarity of house-sites is the separation of manioc cultivation, which is carried on around the house, and coca cultivation, which is done in a separate plot away from the house itself.

... no greater gap

Within these Indian houses, a most disconcerting cleanliness reigns. But the great open yard around the building is lacking, and the jungle encroaches almost directly upon the house from all sides.

Theodor Koch-Grünberg, *Zwei Jahre unter den Indianern* (1923)

Makuna house
Rio Popeyacá, Amazonas

This interesting style of house is confined to the Yukunas of the Rio Miritiparaná, the Tanimukas of the Guacayá and the Makunas of the Popeyaca and lowermost Piraparaná. The now extinct Yaunas of the Apaporis also used this type of construction. Such a geographically restricted range is hard to explain, for the building is one of the most efficient in the entire Amazonia. The roof is made usually of the very durable leaves of a small palm *(Lepidocaryum tenue)* that grows only in sandy or rocky savannah-forests. The vent-holes for smoke and light, though enormous, are always so situated in relation to the generally prevailing winds that the wildest tempest does not drive rain into the house. Four or five families may live together in these houses, some of which have a diameter of one hundred feet or more. Living goes on around the inner periphery, the hard-packed, flat central part being kept free for tribal dances and councils. The circular houses are always surrounded by a circle of cultivated peach-palm trees *(Guilielma speciosa),* and ancient house sites, long abandoned, can be located easily from the air by the circle of these trees which do not grow in the wild state.

... and the jungle encroaches ...

The gradually sloping roof arises from a low wall built in a circle. The roof is crowned in front and in back by a wonderful gable-shaped wide opening smoke-hole which serves also as a window for light and brightens up the interior of the house.

Theodor Koch-Grünberg, *Zwei Jahre unter den Indianern* (1909)

Makuna thatching
Rio Apaporis, Vaupés

Throughout the Colombian Amazonia, Indian houses are usually large structures. They must be built by tribal or communal effort. The circular constructions of the Makunas, Tanimukas and Yukunas are as massive as they are ingenious. The tiny, bothersome gnats so often found in the vicinity of rapids, where these people like to raise their dwellings, are kept from the houses by the dimness and coolness of the interiors. Windows there are none, and the two low doors are usually kept covered. Light, however, does enter from the two ingeniously directed smoke-holes, but the light is so channeled into the house that it brightens only the centre of the interior, not the inner periphery where hammocks are slung, cooking is done and life is carried on by the numerous families inhabiting the roomless building. Some of these circular houses may have a diameter of 75 or 100 feet.

... wonderful gable-shaped smoke-hole ...

The whole facial expression of the women and girls had something melancholy, but infinitely gentle, about it.

Richard Schomburgk, *Travels in British Guiana* (1922)

Kubeo girl
Rio Vaupés, Vaupés

The life of the Indian woman would not seem, from our standards, to be a happy existence. Most of the manual labour of housekeeping and agriculture is left to her. She usually does not take part in any of the decisions concerning tribal economy or politics, and her place in tribal festivities is rigidly circumscribed. Yet she seldom shows exasperation with her lot in life, and one of the first impressions that the traveller amongst these peoples gets is one of how efficient and how contented is the female element in this primitive society.

... melancholy ... but infinitely gentle ...

It is strange that the children are usually weaned only in their third or fourth year, so that the elder often stands quietly in front of its mother, and takes its accustomed nourishment from the one breast while the younger in its mother's arms is sucking at the other.

Richard Schomburgk, *Travels in British Guiana* (1922)

Kubeo Indian mother feeding a three-year-old child
Near Mitú, Rio Vaupés, Vaupés

One of the reasons why mothers breast-feed their offspring for such apparently long periods is the poverty in the Amazonian diet of calcium and other essential minerals for the sustenance of developing young bodies. Chickens often lay eggs without the hard, calcarious shell; and the sparsity of calcium available is attested by the method of preparing the coca-powder with its alkaline admixture -- the ashes of a potassium-rich leaf. Cow's milk and other sources of calcium-rich foods are not easily available.

... weaned only in the third or fourth year ...

They cleanse their bodies and wash their clothing with bark which, beaten in water, produces a thick foam like soap.

Theodor Koch-Grünberg, *Zwei Jahre unter den Indianern* (1910)

Wash day
Rio Kuduyarí, Vaupés

Indians in the Colombian Amazon are invariably clean. Wherever clothes are worn, soap is one of the elements from outside which is traded in through missionary and commercial travellers and rubber workers. It is, however, a very costly item in most of the far off corners.

Before he learned to wear clothing, the Indian was at least free from the worry of procuring soap. He utilized certain plants, containing saponines which could be lathered, to cleanse his skin; but these are not wholly satisfactory for removing grime from cheap cotton goods. It is worthy of note that the naked Indian rarely suffers from skin eruptions and infections, whereas the number of natives -- usually those living close to white man's settlements -- who, through the use of unwashed clothing when soap is unavailable, have skin troubles is unhappily too high. It is my firm conviction that the acquisition of clothing has been a bane instead of a blessing to the Amazonian Indian.

... a thick foam like soap ...

Those [Mirañas whom] I saw had small, thickset strong
and dark bodies with broad faces devoid of any pleasant
expression.

Carl F. P. von Marius, *Zur Etnologie
Amerika's Zumal Brasiliens* (1867)

*Miraña boy
Rio Caquetá, Amazonas*

Today, the once-proud race of the Mirañas is nearly gone. There are
fewer than 80 living along the lower part of the Caquetá and Kahuinarí
Rivers in Colombia and a few in adjacent Brazil. A century ago, they
were many and fierce -- the terror of neighbouring tribes of Indians
towards whom they kept a haughty aloofness. There is still much of
this spirit amongst the few remaining members of this tribe.
Notwithstanding their innate sullenness, they are loyal and
trustworthy and as guides and woodsmen cannot be matched.

... broad faces devoid of any pleasant expression.

The men of the Makuna tribe are splendid, with well-proportioned bodies and pleasing features.

Theodor Koch-Grünberg, *Zwei Jahre unter den Indianern* (1923)

Makuna boys
Rio Piraparaná, Vaupés

Far from the centres of civilisation, relying upon their ingenuity and environment for all the needs of life, the Makunas are an unusually stout and healthy people. Unlike many tribes, the Makunas are increasing in number. Intelligent, loyal and strong, they have successfully defended their country from infiltration by white men of commercial pursuits, earning a fearful reputation as ''wild savages'' by their frequent killings in recent years. They are not treacherous, but, when called upon to match the treachery and guile of an unwanted intruder, they are equal to the task of defending their homes and families with treachery.

... well-proportioned bodies and pleasing features.

Simple as the game was, it could not be denied that it
must contribute a good deal to the adroitness of the limbs
and strengthening of the muscular powers ...

Richard Schomburgk, *Travels in British
Guiana* (1922)

*Yukuna jousters
Rio Miritiparaná, Amazonas*

When tribes come together for the frequent dances concerned usually
with the harvest of sundry fruits, many are the feats of strength and
prowess that take up the time of the younger men. Amongst the
Yukunas, a form of ''wrestling'' is the favourite sport. This consists
in jousting with knotted clubs. Smart and telling blows with the
gnarled protuberances at critical locations on the muscles are the aims
of the contestants. The Yukunas, who are excessive users of coca, are
noted amongst the Indians for their fine physiques and their strength.

... adroitness of the limbs and strengthening of the muscular powers ...

And then, after the lapse of another age, the Red Man
will have gone, and nothing will be left of his transit but
-- a few broken pots.

Franz Keller, *The Amazon and Madeira
Rivers* (1875)

*Barasana father and son
Rio Piraparaná, Vaupés*

In general, the Indians beyond the reach of white man and his culture
are happy. They may be poor and beset with sickness and infirmities,
but they still seem to be happy. This changes when European
civilisation spreads its brittle veneer over these people. Many are those
who willingly accept the new way, but all realise that it does
something to their age-old unity that is irresistible and usually fatal.
The naturalist yearns for the preservation of these people and hopes
that the northwest Amazon, defended as it is by nature from easy and
rapid influx of commercial interests, will not be spoiled as have been
many other primitive and wild corners of tropical areas.

... and nothing will be left of his transit ...

The pride of the woman consists mainly in the possession of a large number of tame domestic animals. What young mammals they can therefore catch, they bring up on the breast, with the result that so great an attachment is implanted in the creatures, especially the monkeys, that they will follow at their foster mother's heels.

Richard Schomburgk, *Travels in British Guiana* (1922)

Tame tapir in Kofán home
Conejo, Rio Sucumbios, Putumayo

The number of animals tamed by the Indians and that live in or near the houses is incredible: various kinds of bird -- parrots, macaws, pajuiles and many tiny birds, several kinds of monkeys, deer, tapirs, and many other animals, including boa constrictors that frequently live in the rafters and keep the houses clear of mice and rats. Most of these tamed animals are captured as very young or recently born offspring or as eggs; in all cases, they are raised in the freedom of the household and live happily with both the human inhabitants and the dogs.

... so great an attachment ...

Very early in life, the children show great intelligence and natural behaviour. To be sure, there are amongst the Indians, just as everywhere else in the world, poorly trained children, especially amongst the very young, who have not yet acquired the calm and self-control of their elders. Then the adults reprove the child with words of admonition. A father might warn a shrieking child who is disturbing the peace and stillness of the night: "Be quiet or the wicked spirit will come!" But I have never seen them carried away to unjust actions and ill-treatment of a child in a fit of sudden wrath.

Theodor Koch-Grünberg, *Zwei Jahre unter den Indianern* (1910)

Tikuna boys
Rio Loretoyacu, Amazonas

Boys from early age seem to live their own lives and to learn in groups of their peers the basic needs of living in their forest environment. Fishing, hunting, house construction and other male occupations they learn from their fathers or older brothers. Their early freedom, however, seems to encourage maturity rather than lead to irresponsibility. At very young ages they exhibit surprising intelligence and ingenuity. The lack of strict disciplinary control apparently has resulted over hundreds of years in an early maturity that might astound parents in our western culture.

... great intelligence and natural behaviour ...

Life is a serious thing to the Indian boy. He is ambitious to become a man as soon as possible. Around him, he sees skillful hunters and fishermen and wants to be like them.

James Rodway, *In the Guiana forests* (1911)

Kabuyarí boy
Rio Kananarí, Vaupés

Indian children are rarely, if ever, punished by their parents, yet there seems to be a deep respect for elders. This respect is especially marked amongst the boys whose keen observation follows every activity of an older brother or father. Sometimes it seems to the casual observer that the boys are given to long periods of moodiness and inactivity, but closer observation shows that the child is intent upon learning some craft or activity and that the "moodiness" is actually deep study. Hours are thus spent by a boy silently watching his father fish, and then one day, without much practice, the boy becomes an expert in his own right. This characteristic of learning by intent observation is one of the traits of the primitive Amazonian which is first to suffer when he comes into sustained contact with white men and their civilisation.

Life is a serious thing to the Indian boy.

How I delight to see those naked boys!
Their well-formed limbs, their bright, smooth, red-brown skin,
And every motion full of grace and health;
And as they run, and race, and shout, and leap,
Or swim and dive beneath the rapid stream
Or, all bareheaded in the noonday sun,
Creep stealthily, with blowpipe or with bow,
To shoot small birds or swiftly gliding fish,
I pity English boys; ...

Alfred Russel Wallace

Makuna fisherboy
Rio Popeyacá, Amazon

The name by which the Makunas designate their tribe is Ee-de-ma-as, meaning ''people of the water.'' From an early age, the boys take to the water as naturally as amphibians. They play and learn to work and fish in the rivers, and boys always accompany older men on the unbelievably long and tedious canoe trips that the elders make to visit distant tribesmen or neighbouring peoples. Provided with an admirable resistance to sun, rain and physical exertion, the men and boys revel in life on the water. On such long peregrinations, the youngest of the boys usually are the fishermen, and the discussion of fishing luck and conditions and jokes and boasts comprise a major part of the almost incessant conversation that accompanies the rhythmic paddling through the daylight hours.

... all bareheaded in the noonday sun ...

All the Indians know the river and its currents.
Downstream the canoe is piloted along the currents that
sweep out from creeks, calling for frequent crossovers;
the centre of the river is usually avoided because of the
sun, the Indians preferring the shade of the shoreline ...
Paddling follows a definite though irregular rhythm. The
prowsman ... is keeper of the stroke ... Unless exhausted
from a long trip the mood of the men in a canoe is gay.
They burst into spontaneous shouts ... and joke loudly.

Irving Goldman, *The Cubeo: Indians of the
Northwest Amazon* (1936)

*Young Kabuyarí paddler
Rio Apaporis, Vaupés*

Boys learn the art of paddling from tender years -- in fact, many
children have tiny canoes in which they learn by mimicking the older
men. Paddling requires great strength and perseverance. Young
children, elderly men and woman usually do not paddle, although
women may, on occasion, have to be at the helm while the strongest
young man serves as prowsman, a position requiring much more
strength and skill than paddling.

All the Indians know the river ...

The spirit of freedom in which the children are brought up is extraordinary. They wander whither they will without parental leave or giving prior notice. They act as they please. They do not mind what their elders ask them to do. Chores they are requested to carry out remain undone .... Such independence it would seem might breed rebellious and difficult offspring, but this does not happen. Offsetting this spirit of autonomy there exists an absolute respect of the tribal laws of discipline, a respect that grows up with the children and that becomes well fixed from the example set by the parents and from ceremonial practices.

Alcionilo Bruzzi Alves da Silva, *Civilização Indígena do Vaupés* (1962)

Young Taiwano naturalist's helper
Rio Kananarí, Vaupés

Most of the young men and boys are serious-minded, inquisitive about everything that goes on and, of course, deeply interested in the animals and plants of their surroundings. They make loyal and hard-working helpers to the naturalist and are genuinely interested in the reasons why a naturalist would come to their region from far away to study and collect material. The wise naturalist will listen to their stories and often learn lessons that might easily be missed had he not taken advantage of the knowledge possessed by the Indians.

The spirit of freedom ... is extraordinary.

What I have called the play pack, the grouping of boys from the age of six to past adolescence, is the major educational institution of the sib. It is in a sense an age grade whose members remain closely associated throughout their lives in most cases.

Irving Goldman, *The Cubeo: Indians of the Northwest Amazon* (1936)

Young Makuna boys in inseparable friendship
Rio Piraparaná, Vaupés

One of the noteworthy customs amongst Indian boys of many tribes of the Colombian Amazon is their close friendship in groups -- groups for play, for paddling, for hunting and fishing with older men of the tribe and even for participation in labour such as thatching a roof. Usually, the members of the group live in one of the communal round-houses and are therefore related by blood or marriage; there are, however, numerous instances when this relationship is not so local.

... the major educational institution ...

If to be a savage means to be rude and uncouth, ill-mannered and disagreeable, then the Indian little deserves such an appellation. He is one of nature's gentlemen, and even when his wishes do not correspond with yours his opposition is only passive.

James Rodway, *In the Guiana Forests* (1911)

Kofán chieftain
Conejo, Rio Sucumbios, Putumayo

It is not at all infrequent to find the leader or chieftain of a tribe to be friendly, helpful, intelligent, trustworthy and dedicated; in fact, to encounter the opposite is indeed a rare experience. The naturalist, interested in plants and animals -- both close to the Indian's preoccupations -- usually is immediately accepted with excessive collaborative attention. These leaders are gentlemen, and all that is required to bring out their gentlemanliness is reciprocal gentlemanliness. Until the frequently unsavoury veneer of western culture surreptitiously introduces the greed, deception and exploitation that so often accompanies the good of ways foreign to these men of the forests, they preserve characteristics that must only be looked upon with envy by modern civilized societies.

... one of nature's gentlemen ...

Among the many activities of a ... shaman, one of the most important refers to his relationship with ... the supernatural Master of Animals. This spirit-being can manifest itself in many guises, but it is generally imagined and seen in hallucinations as a red dwarf ... in the attire of a hunter armed with bow and arrow. He is the owner and protector of all animals ...

Gerardo Reichel-Dolmatoff, *The Shaman and the Jaguar* (1975)

Kofán medicine-man with bamboo lance
Rio Sucumbios, Putumayo

The medicine-men of the Kofán tribe who are experts in the preparation of curare are also usually -- until age overtakes them -- specialists in hunting, teaching the knowledge necessary to younger and inexperienced hunters to lairs of certain animals, their cleverness in avoiding overtaking, the appropriate type of arrow poison to use, the essential food and sexual taboos prior to hunting expeditions and other information considered vital to the chase.

... relationship with the supernatural ...

The chief, in full regalia, was up and about by earliest
sunrise.

Richard Schomburgk, *Travels in British
Guiana* (1922)

*Yukuna chieftain
Rio Miritiparaná, Amazonas*

In most tribes of the Colombian Amazonia, it is the medicine-man
who holds the power of government. The Yukunas, however, have
carefully divided the domains of their medicine-men and their
chieftains. The Arawak people, once a large and warlike group, look
for civil and military leadership to their chieftain, whose emblem of
authority is his long rattling wand with which he opens all conclaves
by rattling the imprisoned pebbles in the fire-swollen end three times
around the outside periphery of the house. Amongst the Yukunas --
and unlike other tribes -- only the civil chieftain may wear boar-tusk
necklaces.

The chief, in full regalia …

The accounts given by the early missionaries of the doings of the payés [medicine-men] are seldom full or reliable. Those pious men regarded them as the great obstacle to the reception of the Christian faith by the natives and always wrote of them with a certain impatience and disgust, under the belief ... that the payés had direct dealings with the devil.

Alfred Russel Wallace (Ed.), Richard Spruce,
Notes of a Botanist on the Amazon and Andes
(1908)

Kamsá medicine-man at work
Sibundoy, Putumayo

Very early in human pre-history certain members of the community acquired a better knowledge of the properties of plants than their fellows and gradually they became leaders and even spiritual powers. The medicine-man has persisted in aboriginal societies everywhere. In the northwest Amazon, they exercise great influence in many aspects of life, not only in the tribes of the humid lowlands but even more so in the groups living in the highlands. In this photograph a widely famous medicine-man is diagnosing a case of illness having taken a drink of the intoxicating floripondio (Brugmansia) to induce the visual hallucinations that will assist him in his efforts.

158

... direct dealings with the devil ...

The office of payé [medicine-man] is not hereditary, but it seems to be fairly common for one of the sons of a well known shaman to follow his father's calling. More important ..., however, are certain psychological and intellectual qualities that mark a person as a potential payé and that will be recognized in his youth ... Among these qualities are a deep interest in myth and tribal tradition, a good memory for reciting long sequences of names and events, a good singing voice and capacity for enduring hours of incantations during sleepless nights preceded by fasting and sexual abstinence .... Above all, a payé's soul should "illuminate"; it should shine with a strong inner light rendering visible all that is in darkness, all that is hidden from ordinary knowledge and reasoning.

> Gerardo Reichel-Dolmatoff, *The Shaman and the Jaguar* (1975)

Kamsá student-apprentice of shamanism
Sibundoy, Putumayo

The young man in this photograph is holding the flower and leaves of the most potent of the South American hallucinogens, known amongst the Kamsá Indians as *culebra borrachera* (Methysticodendron). He will prepare a tea of this plant material, rich in scopolamine and other intoxicating alkaloids, which he will drink in order to induce the necessary psychic and physical condition to see visions and to establish "learning spirit" with his teacher, an elderly and highly respected medicine-man of the tribe.

160

... rendering visible all that is in darkness ...

Shamanism is well developed among the Indians of the Vaupés, and the shaman (or payé, as he is commonly called in that area), is probably the most important specialist within the native culture. It is he who, representing his local group, establishes contact with the supernatural powers and who, to the mind of his people, has the necessary esoteric knowledge to use this contact for the benefit of society.

Gerardo Reichel-Dolmatoff, *The Shaman and the Jaguar* (1975)

Elderly Yukuna medicine-man
Caño Guacayá, Rio Miritiparaná, Amazonas

The work of a medicine-man is not easy: it is constant and physically demanding. It is surprising, therefore, to find many of these practitioners to be men of advanced age. Medicine-men are extremely powerful amongst the Yukunas and Matapies living in the headwaters of the Rio Miritiparaná. In this region they not only act as physicians but exercise their training and knowledge to safeguard members of the tribe who undertake long trips by canoe, to protect crops from all sorts of damage, to control the weather, to converse with ancestors and to accomplish a myriad of other similarly esoteric activities.

… he establishes contact with the supernatural powers …

She had the reputation of being a witch ... and I found, on talking with her that she prided herself on her knowledge of the black art ... I was much amused at the accounts she gave of the place. Her solitary life and the gloom of the woods seemed to have filled her with superstitious fancies ... The witchcraft of poor Cecilia ... consisted of throwing pinches of powdered bark of a certain tree and other substances into the fire whilst muttering a spell ... and adding the name of the person on whom she wished the incantation to operate.

Henry Walter Bates, *The Naturalist on the River Amazon* (1863)

Old Tikuna hag
Rio Amacayacu, Amazonas.

Indian women age early. This is a result of a combination of poor diet, frequent exposure to the elements, overwork and lack of proper care in pregnancy and childbirth. The woman usually performs the heavy tasks of planting and harvesting, household chores, cutting of kindling and all similar types of labour. The felling of forests for agricultural plots or for house raising, hunting and fishing fall to the menfolk.

The weeks of preparation demanded by *cachirís* or tribal festivals and dances are times that tax the utmost energies of all the women. They nonetheless enter into the jollification with enthusiasm, both young and, like this ancient Tikuna woman, old, and dance with a remarkable show of stamina until weariness or drunkenness, or both, puts an end to their exertions and brings on a deep and well won sleep.

Some elderly women, like this old hag, become very authoritative in the tribe by acquiring the reputation of being able to perform hexes. In this way, they hold a kind of local sway, exercising what amounts to a kind of blackmail, inasmuch as the younger members of the group come to fear their presumed knowledge of witchcraft and toxic plants.

... the reputation of being a witch ...

It happened in the beginning of time. In the beginning of time, when the Anaconda-canoe was ascending the rivers to settle mankind all over the land, there appeared the Yajé Woman. The canoe had arrived at ... the House of the Waters, and the men were sitting in the first *maloca,* and there she gave birth to her child ... While the men were preparing *cashirí,* the woman left the *maloca* and gave birth to the yajé vine in the form of a child ... The men were drinking when she had her child and at once they became dizzy ... Only one of them resisted and took hold of the first branch of yajé ... He took off one of his copper earrings and broke it in half and with the sharp edge he cut the umbilical cord. He cut off a large piece. This is why yajé comes in the shape of a vine.

Translation of a Desano Indian myth. Gerardo Reichel-Domatoff, *The Shaman and the Jaguar* (1975)

Anaconda snake
Rio Macaya, Vaupés

The anaconda, greatly feared by readers of travel-books that exaggerate its length and stealth, is very common in the almost uninhabited rivers and tributaries of the northwest Amazon -- so common and abundant that this snake figures in many of the origin myths of the Indians. Although we have known cases where the anaconda has made off with a very young child left alone for a few moments during bathing, this unoffensive snake, its size notwithstanding, must not be numbered amongst the ''dangers'' of the Amazon.

... in the beginning of time when the anaconda-canoe was ascending the rivers ...

The Indians had some special name for, or some special legend about, every peculiarly shaped rock, of which many measured several hundred cubic feet.

Richard Schomburgk, *Travels in British Guiana* (1922)

God-face in cliff
Jirijirimo, Rio Apaporis, Vaupés

The number of Indian legends connected with curiously eroded stones that resemble human or animal figures is myriad. One of the most perfect of these is the face of an Indian god at the Falls of Jirijirimo. According to Taiwano legend, he guards the spirits of the dead chieftains which are represented in tangible form by the huge conglomerate boulders scattered below the waterfall. No Taiwano, save a medicine-man, dares to paddle through the long and narrow chasm downstream from the great falls; he will walk overland, dragging his little dugout over the savannah and through the forest, to a point far below the chasm. The medicine-men have prohibited Taiwanos to look upon this face. All of them know of its existence and will tell the explorer of its fearsome power, but they keep the admonition never to invade the sacredness of the chasm.

... legend about every peculiarly shaped rock ...

Dumb with astonishment, and enthralled at its terribly sublime aspect, we gazed upon the tumult of the struggling waters, the deafening thunder of which swallowed up every other sound.

Richard Schomburgk, *Travels in British Guiana* (1922)

Makunas reciting legend of a waterfall
Rio Apaporis, Vaupés

Most Indians living in the rapid-choked rivers of Amazonian Colombia have legends concerning the spirit or spirits that reside in each of the rapids or falls. To the Indian, it is obvious that only some supernatural being could have thrown up such tremendous obstructions as to keep the waters in a fury throughout all time. The inquiring explorer living and travelling with these people does well to bear patiently with them and listen when, after passing canoes and cargo, they return to some vantage point and recite a story telling of a spirit's heroic exploit, rascality or trickery resulting in the creation of the falls. These stories may be short or they may take hours in telling. The Makunas, a Tukanoan people, believe that many of their falls sprang from a noisy contest, still going on, between an evil demon, usually attributed to an animal, and a good spirit, generally in a plant. Other tribes frequently ascribe the causative contest to enemy constellations of stars.

... the tumult of the struggling waters ...

... this liana has the property, perhaps chiefly antibilious,
of curing the malarial fevers of this region ...

Guillermo Klug, Unpublished notes preserved
in the United States Herbarium, Smithsonian
Institution.

Trunk of the yoco liana
Mocoa, Putumayo

The bark of the *yoco* plant -- *Paullinia Yoco* -- has a caffeine content
of nearly three percent. It is the only caffeine-rich plant the bark of
which is used as a stimulant. It is the morning stimulant of the tribes
of the westernmost Colombian and Ecuadorian Amazonian region: the
Kofáns, Sionas, Inganos and other tribes. All other caffeine-rich
plants that are employed as stimulants have the caffeine concentration
in the seeds, fruits or leaves. The Indians who use yoco take nothing
to eat until nearly noon, having drunk several gourdfuls of yoco at six
in the morning. The effects of the stimulation are strongly felt within
ten minutes following ingestion of the cold-water infusion of the bark.

... the property of curing the malarial fevers of this region ...

They take it to acquire strength, vigour and agility for their long canoe-paddlings, for hunting trips and for their tiring trips through the jungles, and in order not to feel hunger.

Rafael Zerda-Bayón, *Informe del Jefe de la Expedición Cient. del Año de 1905-1906. Lista de Productos del Caquetá* (1906)

Kofán medicine-man rasping yoco Conejo, Rio Sucumbios, Putumayo

To prepare the stimulating yoco drink, the bark of the stems of the liana must be rasped, and the resulting material must be allowed to remain in cold water after considerable kneading of the rasped material to extract the active principle, caffeine.

... to aquire strength ...

Lianas, which have a stout stem at least three inches in diameter at the base, are utilized. Starting at the root, the stem is cut into pieces which may vary from one to three feet in length. These pieces are stored in cool corners of the Indian houses and retain their stimulating properties for a month or even longer.

Richard Evans Schultes, *Botanical Museum Leaflets* Harvard University (1942)

Pieces of yoco stems in Kofán house
Rio Sucumbios, Putumayo

Every Kofán house keeps a store of stems of the yoco liana for daily use as "breakfast" or for long canoe or hunting trips. The stimulant effectiveness of the plant will last for a month or longer. The pieces of the trunk are kept in the coolest part of the house. On trips of four or five days duration, the Kofáns take no food but exist on the highly stimulating effects of yoco.

... retain their stimulating properties ...

... to foresee and to answer accurately in difficult cases, be it to reply opportunely to ambassadors from other tribes in a question of war; to decipher plans of the enemy through the medium of this magic drink and take proper steps for attack and defense; to ascertain, when a relative is sick, what sorcerer has put on the hex; to carry out a friendly visit to other tribes; to welcome foreign travellers; or, at last, to make sure of the love of their womenfolk.

Manuel Villavicencio, *Geografía de la República del Ecuador* (1858)

*A young liana of caapi cultivated in the forest
Rio Popeyacá, Amazonas*

Although the Indians prefer the older and more robust lianas of the caapi plant, they occasionally have the plant cultivated in their agricultural plots or even in the forests. They maintain that the hallucinogenic drink prepared from the bark is much stronger and longer lasting in its effects when prepared from the older lianas. The vine *(Banisteriopsis Caapi)* has numerous names in the northwest Amazon: *caapi* in the Vaupés of Colombia and adjacent Brazil; *ayhuasca* in Peru and Ecuador; *yajé* in the westernmost parts of the Colombian Amazon. The drink basically prepared from this plant is sometimes fortified by the addition of leaves of another vine of the same family *(Diplopteris Cabrerana)* or from a small tree of the Coffee Family *(Psychotria viridis)*. These additives, the Indians say, strengthen and lengthen the intoxication, an assertion that has been substantiated by chemical analysis of the several plants involved.

... foresee and to answer accurately in difficult cases ...

They add to the yajé the leaves and the young shoots of the branches of the oco-yajé or chagro-panga ... and it is the addition of this plant which produces the "bluish aureole" of their visions. These are like cinematographic views and occur after about a half litre of the drink has been consumed ... Thereafter, the Indian falls into a profound sleep ... During this period the subconscious activity acquires enormous intensity. The dreams follow each other with extraordinary precision ... giving the power of double vision and of seeing things at a distance ... Upon awakening, he retains clearly the hallucinations ... which he experienced in unknown regions.

Guillermo Klug, quoted in C. V. Morton, *Journal of the Washington Academy of Science* (1931)

Leaves of the oco-yajé
Mocoa, Putumayo

In the vast area of the western Amazonia where ayahuasca or caapi is used as a sacred hallucinogen, there are many plant additives put with the drink made basically from *Banisteriopsis Caapi*. There are two, however, that are very widely employed: *chacruna,* the leaves of a Psychotria of the madder family; and *oco-yajé* or *chagro-panga,* the leaves of *Diplopteris Cabrerana* of the same family as Banisteriopsis. Both of these plants have in their leaves a different hallucinogenic chemical -- a tryptamine -- from the active principles -- beta-carbolines -- found in Banisteriopsis. Addition of these leaves greatly enhances and lengthens the intoxication through a kind of synergistic activity. How, one wonders, did these Indians find from the 80,000 species around them these two additives with such extraordinary effects?

... this plant ... produces the "bluish aureole" of their visions.

There was a woman among them [the first Tukano men], the first woman in Creation, and while the men's excitement was growing and the house began to be filled with the voices and movements of the crowd, she left it unobserved and went outside. The woman was with child. When the Sun Father had created her in the House of the Waters, he had pregnated her body through the eye; by looking at his radiance she had become pregnant and now she was about to give birth to a male child, a child that was going to be *yajé*, the narcotic vine, a superhuman child that was born in a blinding flash of light.

Gerardo Reichel-Dolmatoff, *Beyond the Milky Way* (1978)

Caapi vine cultivated in Fusugasugá
Original plant from Mocoa, Putumayo

The Indians prefer to make their hallucinogenic drink from old lianas growing wild in the forest, although they do cultivate the plant in their gardens. They maintain that the older, more extensive and thicker stems yield a stronger narcotic bark than the slender cultivated vines. The illustration shows how heavy a liana this plant, *Banisteriopsis Caapi,* will grow in time in a botanical garden in Fusugasugá.

182

… she gave birth to a male child, a child that was going to be yajé, the narcotic vine …

The [caapi] drink is boiled in any container but is later put into another special, sacred pot ... In the manufacture of this pot, made by elderly women, a particular very hard, smooth, yellow stone is used ... to make the clay compact and polish the surface of the pot before firing it ... As with other ceremonial objects, this pot comprises three parts: the "yellow" base; the "red" body and the "blue" upper part, since in serving the drink communication with the supernatural world is established. Actually, the pot is identified with the Snake-canoe ... and, since man arrived in the world [from the Milky Way] in the Snake-canoe, so now the return journey begins [with the help of the hallucinogen].

> Gerardo Reichel-Dolmatoff, *Desano --*
> *Simbolismo de los Indios Tukano del Vaupés*
> (1968)

Ceramic caapi pot
Jinogojé, Rio Apaporis, Vaupés

The caapi pot in which the hallucinogenic drink is kept during the ceremonies is a very special possession of every large round-house or *maloca*. Following its use, the pot is carefully cleaned and hung up always outside the house in the eastern corner of the overhanging roof. The manufacture of these pots involves specific ritualistic incantations. They are always coloured in black and red, and the designs are those suggested by the geometric figures experienced during the caapi-intoxication. Some of these pots are the property of several generations and are passed down according to rigid tribal tradition.

... identified with the Snake-canoe ...

It was dark now, and one of them lit the *turí,* the large resin-covered torch standing near the centre of the room, and it began to shed an intensive red light over the scene. There was a sudden hush.

Gerardo Reichel-Dolmatoff, *The Shaman and the Jaguar* (1975)

Makuna medicine-man starting the caapi-ceremony
Rio Popeyaká, Amazonas

Stems of the caapi-vine are collected in the early afternoon. Upon returning, the pieces of vine are macerated with a hard wooden pestle during which there is constant chanting, presumably recounting the legendary story of the discovery of caapi. Another Indian took down the painted ceramic caapi-pot that is kept hanging outside the house; he cleaned it with a heavy feather, for it is never washed. Then the macerated stems were put into cold water in the pot and allowed to stand for several hours until the liquid took on the colour of milk. It was then passed through a sieve.

By this time, night had fallen, and the men went into the maloca to dress in their feather ornaments and fit rattles to their elbows and ankles for the dance. At this time, the head man of the festival lighted the pitch-torch and chanted for at least twenty minutes. His wand of authority was a long stick, swollen by fire and slit at the end into which pebbles had been placed. During the chanting, he walked around the house violently shaking his wand. He finally took his place with the other men and the pipes-of-pan began to play, whilst the caapi-pot was passed around, each man taking the first of many draughts of the hallucinogenic drink.

... the *turí* shed an intensive red light ...

... by taking copious draughts of caapi, both physician-shaman and the sick one would be visited by hallucinations. The terror of the unknown makes the mind of man more submissive to mystic influences, and so dreams, brought on by caapi, the vine-of-the-souls, establishes a contact, in their minds, between the dead, who must have the answer to the sickness, and the living who have need of it.

Victor W. Von Hagen, *South America Called Them* (1949)

Makuna medicine-man and his student
Rio Piraparaná, Vaupés

It is a firm belief amongst Indians of the northwest Amazon that all sickness and even death is the result of the machinations of malevolent spirits from outer realms. Consequently, the medicine-man, through various hallucinations, can diagnose the causes as a result of communication with these spiritual forces.

One of the most widely employed hallucinogens in the northwest Amazon is caapi or yajé. In this photograph, the medicine-man with his student, both intoxicated with caapi, are dancing around the prone figure of a young man, probably an epileptic, in an attempt to discover the cause of his abnormal condition.

... a contact between the dead, who must have the answer to the sickness, and the living ...

All Tukano payés take *viho* snuff prepared from Virola sap, and any man may occasionally take the snuff, usually under the guidance of an experienced payé. The trance produced by the narcotic is said to be the most important means by which one can establish contact with the supernatural sphere in order to consult the spirit beings, above all Viho-mahse, ''snuff-person'' (the owner and Master of Snuff), who dwells on the Milky Way whence he watches continuously the doings of mankind.

Gerardo Reichel-Dolmatoff, *The Shaman and the Jaguar* (1975)

The bark from which the potent viho snuff is prepared
Rio Piraparaná, Vaupés

In Colombia, the use of the snuff prepared from the resin-like exudate of the bark of several species of Virola (especially *V. theiodora),* is usually restricted to medicine-men. Amongst the Waika in adjacent Venezuela and Brazil, however, use of this hallucinogenic snuff is general amongst all males above sixteen years of age. In the Colombian and Peruvian Amazon, the resinous exudate is prepared, not as a snuff, but in small pills that are ingested. The active principles are tryptamines in both cases.

... to consult the spirit beings ...

The most outstanding example [of native decoration] are the paintings executed on the front walls of the malocas. These walls, made of large pieces of flattened bark, are covered wholly or in part with bold designs painted with mineral colours -- occasionally with the admixture of vegetable dyes -- and represent ... large eye-shaped designs, the wavy lines of the Anaconda-canoe, the *vahsu* [a rubber tree] design, and the design for exogamy ... the Master of Game Animals or individual animals such as fish, frogs or snakes. When asked about these paintings the Indians simply reply: ''This is what we see when we drink yajé; they are *gahpi ohori.*''

Gerardo Reichel-Dolmatoff, *The Shaman and the Jaguar* (1975)

Barasana round-house
Rio Piraparaná, Vaupés

Many -- in fact most -- of the paintings on the front of the Indian malocas in the Rio Piraparaná and other regions are designs seen by the natives during their visual hallucinations whilst under the influence of the sacred narcotic caapi (sometimes referred to as yajé). In the Indian belief, these paintings, representing something that has come to their conscience from the outer, supernatural realm, may protect the house and its inhabitants from danger and disaster.

This is what we see when we drink yajé …

... the various forms of shamanism practiced today ... with the aid of tobacco occupy a central position in tribal culture. They seem to me to constitute true survivals of a more ancient shamanistic stratum with roots in Mesolithic and even Paleolithic Asia, introduced into the Americas 15,000 to 20,000 or even more years ago. Although attenuated and certainly overlaid with more recent features ... they seem to belong to... an archaic shamanistic substratum underlying and to some extent uniting all or most aboriginal American Indian cultures.

> Johannes Wilbert, ''Tobacco among the Warrao of Venezuela'' in Peter T. Furst (Ed.), *Flesh of the Gods -- the Ritual Use of Hallucinogens* (1972)

Yukunas taking snuff
Caño Guacayá, Rio Miritiparaná, Amazonas

Snuffing is one of the commonest methods of using tobacco amongst most of the tribes of the Vaupés and Caquetá river basins in Colombia. The dried and finely pulverised leaves of tobacco are mixed in about equal quantity with finely powdered ashes of the leaves of the yam. The resulting snuff is greyish, not brown, and is pleasant, not too strong. It is always taken through hollow bird-bone snuffing tubes, either in U-shape for self-administration or straight for blowing the powder into the nostrils of a friend. The dose is large, amounting usually to about a teaspoonful for each nostril. Most of the tribes using tobacco-snuff keep the powder in a large land-snail case which is fitted at the top with a short bird-bone tube fixed to the shell with pitch. Men only take snuff. Although the habit is common amongst the Kubeo and Tukanoan peoples of the Rio Vaupés itself, the tribes which are most habituated to snuffing are those living on the affluents of the lower Apaporis (the Popeyacá and Piraparaná) and the Yukunas and Tanimukas of the Rios Miritiparaná and Guacayá.

... survivals of a more ancient shamanistic stratum ...

The tobacco was possibly the first narcotic ever used in South America. In one form or another, it is a prime ingredient in the medicine of the payés.

Alfred Russel Wallace (Ed.), Richard Spruce,
Notes of a Botanist on the Amazon and Andes
(1908)

*Young student payé of the Yukuna tribe
Caño Guacayá, Rio Miritiparaná, Amazonas*

The payés or medicine-men in the northwest Amazon either smoke or snuff tobacco in their treatment and diagnosis of disease. Tobacco is seldom smoked in the Colombian Amazon by unacculturated Indians, but the use of snuff is the normal methods of use.

Medicine-men usually have their understudies or apprentices who start as young men to learn the trade. This photograph depicts a young Yukuna student who is about to take snuff in a hollow bird-bone snuffing tube, measuring out the snuff from a large snail shell in which it is kept. He is dressed in the costume worn during the annual Kai-ya-ree Dance of this tribe.

... a prime ingredient in the medicine of the payés.

Behold how thick with Leaves it is beset;
Each leaf is Fruit, and such substantial Fare,
No Fruit beside to rival it will dare ...
Our Varicocha first this Coca sent,
Endow'd with Leaves of wondrous Nourishment,
Whose Juice succ'd in, and to the Stomach tak'n
Long Hunger and long Labour can sustain;
From which our faint and weary Bodies find
More Succour, more they cheer the drooping Mind,
Than can your Bacchus and your Ceres joined.

Abraham Cowley (1662) quoted by
W. Golden Mortimer, *History of Coca* (1901)

Makuna Indians harvesting leaves of coca
Rio Popeyacá, Amazonas

Most agricultural work is done by the women. The care of the coca plant, from its planting to the gathering of the leaves, however, devolves wholly on the men. Furthermore, coca is planted in a special plot, not with the yuca (tapioca) and other food crops. This isolation of coca and its male-oriented care, together with its importance in origin myths, may be interpreted as indicative of great antiquity of this semi-sacred plant that still occupies a major role in most tribes of the northwest Amazon.

... each leaf is fruit ...

The Indians in the forest regions, particularly those of the Putumayo, toast the leaves and mix them with wood ash or the roasted leaves of a kind of sycamore or setico tree. The whole is then pulverized with a large pestle and mortar .... As required, a quantity is placed in the mouth .... The saliva when swallowed causes the user to lose all sense of hunger .... I have used it on many occasions during my wanderings and find that it is a great help and does not appear to have any ill effect.

Joseph F. Woodroffe, *The Upper Reaches of the Amazon* (1914)

Kubeo woman toasting leaves of coca
Rio Kuduyarí, Vaupés

When the men collect coca leaves during the afternoon and bring them back to the round-house, they are gently toasted in order to be pulverised. This toasting is the work of either men or women. Coca powder must be made each day or, at the most every other day, for it rapidly loses its special taste and, the Indians all maintain, its potency decreases. The collection and preparation of coca occupies a major part of the daily life of those tribes that use this narcotic plant.

In the lower right of the photograph can be seen the leaves of the species of Cecropia *(C. sciadophila)* or *guarumo* that are burned for the ashes which are mixed with the powdered coca as a necessary alkaline admixture.

... to lose all sense of hunger ...

... the coca leaves ... had been baked in a large earthenware pot over a hot fire, all the while being stirred with a hooped stick. The toasted leaves were then powdered inside a short wooden cylinder.

Brian Moser and Donald Taylor, *The Cocaine Eaters* (1965)

Kubeo Indian preparing to pulverise coca leaves
Río Kuduyarí, Vaupés

In the northwest Amazon, coca leaves are powdered and mixed with the alkaline ashes of the leaves of a species of *guarumo (Cecropia sciadophylla)*. Coca powder must be prepared fresh each day in order to keep its taste and strength. The leaves are collected in the afternoon and immediately toasted; they are then reduced to a powder in a hollow mortar made from a tree trunk. The regular thumping of the pestle -- a stick made from a hard wood -- is one of the enjoyable sounds in the great round-house in the early evening. It may often be accompanied by chanting, the legendary stories of how the tribe acquired coca, done by one or two of the older men of the group. It is in this way that the ancient beliefs of the tribe are learned by the young.

... the leaves had been baked over a hot fire ...

The leaves are roasted and then pounded in a mortar made
of the trunk of the pupunha palm, from four to six feet
long, the root being left on for the bottom and the soft
inside scooped out. It is made so long on account of the
impalpable nature of the powder, which would otherwise
fly up and choke the operator.... The pestle is made of any
hard wood.

Alfred Russel Wallace (Ed.), Richard Spruce,
Notes of a Botanist on the Amazon and Andes
(1908)

Tukano Indian pounding dried coca leaves
Rio Vaupés, Vaupés

Coca powder must be made every evening to be fresh. After the leaves
are toasted, they are pulverized in a huge, hollowed out mortar with
a pestle made of a very hard wood. Unless enormous quantities of the
coca powder are to be prepared for a fiesta, this work is done after dark
in the large round-houses. One of the typical and pleasing sounds in
these houses is the regular, subdued rhythm of the thumping of the
pestle -- a sound that may last up to an hour or more immediately after
sundown.

... the impalpable nature of the powder ...

So intimately entwined is the story of coca with these early associations -- with religious rites, with superstitious reverence, with false assertions and modern doubts -- that to unravel it is like to the disentanglement of a tropical vine in the primitive jungles of its native home.

W. Golden Mortimer, *History of Coca* (1901)

Barasana taking coca
Rio Piraparaná, Vaupés

The powdered coca-ash mixture, which is used by the Indians of Colombia's Amazonian regions, is usually taken from a gourd by spoons of tapir-bone. In the uppermost reaches of the Piraparaná, however, the Barasanas and Makunas employ an ingenious little bag of powdered bark tied at the mouth from which protrudes a hollow bird-bone tube. The powder is injected into the mouth simply by vigourous squeezing of the bag. During the day, the men hang their coca-bags from a fibre cord belt which supports the loin-cloth or *guayuco*. I have never seen this apparatus amongst any other Indians.

... like to the disentanglement of a tropical vine ...

Curare, which is the most deadly poison known to the Indians, is prepared by relatively few tribes, though it is traded throughout wide regions.

> Alfred Métraux in J. H. Steward (Ed.),
> *Handbook of South American Indians* (1949)

Kofán medicine man preparing curare
Conejo, Rio Sucumbios, Putumayo

While the use of curare on darts and arrows is slowly dying out in some parts of the northwest Amazon as a result of the introduction of shot-guns, a number of tribes still proudly prepare various kinds of arrow poisons. One of the groups where knowledge of plants involved in the manufacture of curares is still intact and where perhaps the most extensive list of plants that enter the preparation of arrow poisons is the Kofán tribe who live in the Colombian and Ecuadorian areas bathed by the Rio Sucumbios.

Of all the tribes in the northwest Amazon, the Kofán seem to prepare the greatest number of different arrow-poisons. They not only employ the widely used species of *Chondrodendron* and *Strychnos,* but many other plants either alone or in complex recipes -- plants unknown as poisons by most other tribes. It is probably no exaggerations to state that the Kofán Indians are the poison-makers *par excellence* in the Colombian Amazon. The curares and other arrow poisons are carefully prepared with the appropriate ceremonies, chants and fasting necessary for the best results. These Indians, furthermore, have many arrow poisons for specific hunting purposes, depending mainly on the kind of animal they want to shoot.

... the most deadly poison known to the Indians ...

The skill with which the Indian uses the blowgun is almost unbelievable.

Gastâo Cruls, *Hileia Amazonica* (1944)

Ingano hunter
Mocoa, Putumayo

The accuracy with which the forest Indians use the blowgun is hard to believe without actually witnessing it. The blowgun, varying from five to nine or ten feet in length, according to the tribe, can be aimed at a small bird in the crown of a hundred-foot tree, and the target is rarely missed. The Ingano Indians of the vicinity of Mocoa in the Comisaría del Putumayo are known far and wide for their deadly accuracy with the blowgun. This may be due in part to the great care and skill which they, and their neighbours, the Kofáns, exercise in the making of the instrument. These Indians use the rough skin of the stinging ray, which abounds in the rivers of the Putumayo, as a sandpaper with which a perfect bore is polished in the two pieces which, when fitted together, are bound around to form the blowgun.

The skill ... is almost unbelievable ...

We soon heard the chattering of monkeys in the tree-tops
and deftly inserting one of the thin poisoned arrows in the
ten-foot tube he pointed the weapon at a swiftly moving
body among the branches, and filling his lungs with air,
let go. With a slight noise, hardly perceptible, the arrow
flew out and pierced the left thigh of a little monkey.
Quick as lightning he inserted another arrow and caught
one of the other monkeys as it was taking a tremendous
leap through the air to a lower branch.

Algot Lange, *In the Amazon Jungle* (1912)

Makú Indian hunters with blowguns
Rio Piraparaná, Vaupés

The Makú Indians have the reputation in the Colombian Amazon as
makers of the most potent curare preparations. While they utilize the
basic ingredients that are used by other tribes, their numerous
additives differ. These semi-nomadic Indians, extremely primitive in
comparison with other groups of the region, have undoubtedly the
most perspicacious knowledge of the properties of plants. They have
no firearms nor do they cultivate plants: they depend wholly on the
collection of wild fruits and on their expert hunting and fishing ability.

With a slight noise ... the arrow flew out and pierced the left thigh of a little monkey.

Handling the blowgun demands great dexterity and much bodily strength .... A strong man can blow the small dart with such power that he is able to hit the target accurately from a distance of 90 to 120 feet and still attain its full effect.

Theodor Koch-Grünberg, *Zwei Jahre unter den Indianern* (1909)

Young Makuna hunter
Rio Piraparaná, Vaupés

Learning to handle the blowgun begins in early childhood when the Indians make small blowguns for the boys to shoot at targets in the vicinity of the great roundhouse. By the time the boys are mature young men, most of them are experienced hunters with this instrument. It requires not only rapidity of action and skill but physical strength. The distances to which an experienced hunter can shoot with accuracy even small animals in motion is uncanny. Even more inexplicable is the Indian's ability to run through the undergrowth of the thick forests without entangling his eight-foot blowgun in the bushes and vines.

... great dexterity and much bodily strength ...

The distal end of the dart is fitted with a spindle-shaped tuft of light-weight kapok fibre sufficiently large to fill the hole of the blowgun tightly enough to provide resistance to the hunter's puff.

Theodor Koch-Grünberg, *Zwei Jahre unter den Indianern* (1909)

Yukuna boy fitting kapok on darts
Caño Guacayá, Rio Miritiparaná, Amazonas

When a good quantity of darts for the blowgun is prepared, the poisonous syrup of curare is applied to the pointed inch or two of the dart; a slight notch above the poison is made in the dart to facilitate its breaking and leaving the toxic tip in the animal's body. The final chore is the application of kapok silk to the opposite end of the dart. This plug of fibrous material provides resistance to the breath of the blower with the resulting expulsion of the dart through the six- or eight-foot blowgun. Once provided with their kapok-plugs, the darts are stored in orderly fashion in a tubular quiver.

... kapok fibre to provide resistence to the hunter's puff ...

The poisoned darts are, as a rule, made of black, heavy palm wood, seldom from white, light-weight wood and are as thick as a strong knitting-needle. They measure about 16 inches in length and have a sharp tip. About an inch and a half of the pointed end is covered with curare, above which space there is a slight notch.

Theodor Koch-Grünberg, *Zwei Jahre unter den Indianern* (1909)

Poison dart-maker of the Kofán tribe
Rio Sucumbios, Putumayo

While the Kofán Indians have special craftsmen to make their famous blowguns, there are particular men in the group who are considered experts in preparing the darts made from the hard wood of certain palm trees. The darts are made perfectly straight, any imperfect specimens being cast aside. This care is responsible in part for the incredible accuracy of the shooting that these Indians experience with the blowgun. When a large supply of these darts is ready, another tribesman, the medicine-man who makes and handles curare, paints the tips of the darts with the poison. A notch is made below the poison to aid in breaking of the dart in the animal's body.

... the darts are made of black, heavy palm wood ...

Their arms consist chiefly of blowguns or *bodoquedas*, although at present shot-guns and machetes are beginning to be introduced among them ... This celebrated weapon is a hollow, tapering pole, from two to four metres long, pierced longitudinally by a hole some three-sixteenths of an inch in diameter. The outside of the pole is wound around with thin strips of tough bark over which is applied a smooth, black coating of gum-resin ... while the thicker end terminates in a mouthpiece into which a small arrow ... is introduced. The mouth is then applied to the mouthpiece, and with the breath the little arrow is shot out with great force ...

W. E. Hardenburg, *The Putumayo -- The Devil's Paradise* (1912)

Kofán Indian making a blowgun
Rio Sucumbios, Putumayo

Amongst the Kofán Indians, who have the reputation of making the best blowguns, the craft is highly respected and the skill is in the hands of special experts. In this part of the Colombian Amazon the Kofáns are rapidly becoming acculturated, but despite the availability of firearms, many Indians still prefer the blowgun for hunting, and Indians of other tribes in the vicinity seek Kofán-made blowguns.

The outside of the pole is wound around with thin strips of tough bark ...

When my Indians returned with the *timbó,* we all set to work beating it on the rocks with hard pieces of wood .... It soon began to produce its effects; small fish jumped up out of the water, turned and twisted about on the surface or even lay on their backs and sides. The Indians were in the stream with baskets hooking out all that came in the way ....

Alfred Russell Wallace *Narrative of Travels on the Amazon and Rio Negro* (1853)

*Kofán fisherman beating stems of timbó
Santa Rosa, Rio Sucumbios, Putumayo*

When enough pieces of the stems of the *timbó* liana (Lonchocarpus) are gathered, the bark is beaten off and macerated. The active principle -- rotenone -- is in the bark. There are other plants, some cultivated, that can be used in fishing, but timbó is the preferred poison. The term *barbasco* includes all ichtyotoxic plants. The two that are commonly cultivated are a member of the daisy family *(Clibadium)* and a bushy species of the spurge family *(Phyllanthus);* these are used when a few fish are needed, but when a tribal festival is planned and there will be many mouths to feed, there is no substitute for timbó. There are also numerous fish poison plants that are wild and, like timbó, must be collected in the forest. In the northwest Amazon, there are more than thirty different plants known to be used as fish poisons.

... beating it on the rocks ...

By far the most wholesome and general way in which fish
are obtained is through the use of poison ... They dam the
stream ... and then throw the mashed barbasco in ... The
fish frequently jump out of the water, gasping as though
they were being strangled, and the Indians secure these
distressed fish in outspread palm leaves.

Thomas Whiffen, *The Northwest Amazons*
(1915)

Kofán fisherman poisoning the water
Conejo, Rio Sucumbios, Putumayo

The Indians of the northwest Amazon have discovered many species
of plants with chemical constituents that can be used to stupefy fish.
Some of the plants have alkaloids, some have glycosidal compounds.
The most commonly employed ichthyotoxic plant is an extensive
forest liana known as *timbó* (species of Lonchocarpus of the bean
family); its active principle is a ketone -- rotenone -- which is used in
our own agriculture as a biodegradable insecticide. The bark of the
liana is scraped off and macerated, then thrown into or dragged
through the still water in a porous bag or basket. Most fish poison
plants affect the gills so that the fish come to the surface stunned and
seeking oxygen. The flesh of the fish is fully edible, since none of the
toxic principles are absorbed.

The fish frequently jump out of the water ...

The forest of the Black Water contains little fish and harbors less game.

Henry Alexander Wickham, *Rough notes*
(1872)

Fish weir
Rio Guainía, Vaupés

It is a fallacy that all Amazonian rivers abound in fish. In many of the clear-water rivers that carry little or no silt, fish are a rarity, and ingenuity is necessary to provide enough for the daily diet. Elaborate fish weirs, such as the one in the photograph, are placed at proper locations and are visited daily by the Indians. Even then, the days marked by a good catch are few and far between. Humboldt and Spruce both spoke of the difficulty of getting fish on the upper Orinoco and along the Rio Negro and its tributaries, and this paucity of fish caused Spruce to suffer many a day of hunger during his years in the region.

... little fish and ... less game.

We had the good fortune ... to overtake an Indian who was visiting his "cakouri", as the fish trap is locally called ... These traps are common to the river and are built in the shape of a triangle of slats of palm ... Securing the fish after they are in the trap is by the very simple process of dropping into the triangle from on top and hand-catching them ... By such traps and by shooting with arrows from small canoes ... the Indians get practically all their fish ....

Caspar Whitney, *The Flowing Road* (1913)

Black water kakurí (fish trap)
Raudal Tatú, Rio Vaupés, Vaupés

In the rivers that carry heavy loads of silt -- the so-called "white water rivers" -- fish are abundant and are easily caught by conventional methods: lines, nets or poisons. The so called "black water rivers," with clear, almost silt-less waters, are notoriously poor in fish fauna. In these latter waters, the natives have employed plant fish poisons in still or very slow moving inlets or, at the base usually of rapids, the *kakurí* or fish trap. These traps are ingenious inventions and supply the population with an abundance of the smaller fish that are swept over the rapids to be trapped in these weirs.

By such traps ... the Indians get practically all their fish ...

The first thing the boy takes to when he arrives at the age of reason, are the bows and arrows made for him by his father or elder brother ... He attains so much vim, vigour and versatility by tumbling about upon the trees, I might almost say by living continually amidst the denizens of the forest, that he can soon accompany his father when on the chase and when catching fish.

Richard Schomburgk, *Travels in British Guiana* (1922)

Makunas shooting fish
Rio Popeyacá, Amazonas

Uncanny in their knowledge of the habits and habitats of the many fish of these rivers, the Makunas or ''people of the water'' would seem to be the best fishermen of all the Indians of the Colombian Amazon. Their pots never want for fish. Large fish are shot with the bow and arrow, but other methods are also employed when conditions call for them. These people fish with hooks of bone, with plant poisons and with traps and weirs. The net, which is apparently unknown to them, would be of little use as the waters of the Apaporis and its affluents, unlike the Amazon itself and other rivers to the south, are not teeming with fish.

... the first thing the boy takes to ...

Indian farming is of the rudest character. The plantation is simply an irregular clearing in the woods with half-burned logs scattered all over the surface ... The ground has not been turned at all. The mandioca-cuttings are simply placed, several together, in holes dug in the unprepared ground, and they get hardly any care. As a matter of course, the top-crust, baking in the sun and drained by the strong-growing plants, is soon exhausted; every four or five years the old clearings are abandoned, and new ones are made, involving fresh destruction of the forest and great outlay of labor.

Herbert H. Smith, *Brazil: the Amazons and the Coast* (1879)

Indian agricultural plots
Rio Piraparaná, Vaupés

There has been much discussion concerning the usual slash-and-burn system of agriculture prevalent in the Amazonian regions: some experts have condemned it, whereas others have indicated that it is the only type that can be supported by the fragile soil of the hot, moist tropical areas of the world. It has recently, however, been adequately established that modern, mechanized agriculture of the kind practised in temperate parts of the world is not appropriate in these warmer areas. And it is true that this Indian system, developed over millennia, is admirably adapted to the aboriginal lack of machinery and that it inflicts far less havoc on the natural vegetation.

Indians normally fell no more area than they need. After a number of years, a new area is cut for planting, since the soil is "worn out" and, in many places, the leaf-cutting ant proliferates to such an extent that much of the planting is destroyed. It takes often 200 years for the climax forest to re-establish itself through a series of successions of different plants. These small areas are capable of re-establishing a climax cover, but when enormous extents are "efficiently" cleared mechanically in the Amazon, the original vegetation can never return, and the area, unless planted to tree crops appropriate for local conditions, will become a permanent scrubby wasteland forever lost to reclamation.

... farming is of the rudest character.

Imagine the trees of a virgin forest cut down so as to fall across each other in every conceivable direction. After lying a few months, they are burnt; the fire, however, only consumes the leaves and fine twigs and branches; all the rest remains entire, but blackened and charred. The mandioca is then planted without any further preparation.

Alfred Russel Wallace, *A Narrative of Travels on the Amazon and Rio Negro* (1889)

Taiwano maloca
Rio Kananarí, Vaupés

In the Colombian Amazon, the Indian likes to clear large areas immediately around his house. Here he sets out his food staple, the manioca plant.

Contrary to common belief, soil in the northwest Amazon is clayey and poor in nutrient minerals. Once the forest cover is removed by felling and the leaves and small branches are burned, the heavy rains leech out from the upper foot or so of soil all soluble mineral nutrients. It is precisely because of this impoverishment that cereal grasses and other annuals with superficial roots refuse to grow or else grow rhachidically with insignificant yields after two or three years. The Indian has learned that he must get his carbohydrates from a deep-rooted perennial capable of penetrating to where the waters cannot wash out the sparse minerals. He has consequently built his whole agriculture around the tapioca plant which he sows by sticking little twigs or cuttings directly into the soil amidst the tangled mass of partly burned vegetation that was felled six months before the planting and allowed to "dry" out during the months of lighter rain which is known euphemistically as the "dry season."

Once the men have felled and burned over a piece of land, the women plant the tapioca plant or cassava. This basic food plant is always grown near the dwellings, whereas the coca fields are separate and usually a good distance from any other cultivation, possibly an indication of long use of coca in the region and its almost semi-sacred role in local society.

234

The mandioca is then planted …

Long ago, the daughter of a chief ... was discovered to be
with child. The chief wished to punish the author of his
daughter's dishonor, but the girl steadfastly declared her
innocence. The chief was deliberating to kill her when a
white man appeared to him in a dream and told him not
to kill the girl because she was really innocent and virgin.
In the fullness of time, the girl gave birth to a beautiful
female child who was white, like the vision ... The child
was called Mani; she walked and talked precociously; but
at the end of a year she died, without being ill or giving
any signs of pain. She was interred in the house ... and
the grave was watered daily. At the end of some time, a
plant sprang out of the grave; and because it was unknown
to them, the people did not pull it up. It grew, and
flourished, and gave fruit. The birds who ate the fruit
became intoxicated, and this phenomenon increased the
superstitious care of the Indians. At length, the earth
cracked open; they dug into it, and in the root which they
found they believed that they saw the body of Mani. They
ate it and thus learned the use of mandioca. Hence the
name *Mani-oca*, the house or hiding place of Mani.

      Herbert H. Smith, *Brazil: The Amazons and
the Coast* (1879)

*A cultivation of yuca or mandioca
Rio Kananarí, Vaupés*

Yuca (Manihot esculenta), or the tapioca plant, is the staple source of
carbohydrate nutrition in the Amazon. There are two kinds of *Manihot
esculenta:* the "sweet" and the "bitter". They are both of the same
species but differ in the distribution and amount of a poisonous
constituent -- a cyanogenic glycoside -- in the root. The "sweet" type
has less of the toxic substance, and it is concentrated in and near the
bark which, of course, is peeled off; the "bitter" type has the
glycoside in greater concentration and distributed throughout the
starchy part of the root. Indians of the Colombian Amazon prefer
almost exclusively the poisonous type. They have developed methods
of removing the poison because of the great nutritional utility of the
root and the adaptability of the plant as an Amazon crop.

    It is significant of the importance of this plant that, according
to Tukano origin legends, the first Tukanos who came down from the
Milky Way in a dug-out canoe drawn by an anaconda snake had,
together with the man and woman, three plants: caapi, coca and yuca.

... they saw the body of Mani.

To eliminate the poison and render it fit for food, the manioc is subjected to several processes ... The peeled roots are washed ... then cut longitudinally into three or four sections which are put into a bowl ... to soak for twenty-four hours.

Thomas Whiffen, *The Northwest Amazons*
(1915)

Taiwano children cutting and peeling yuca roots
Rio Kananarí, Vaupés

The women till the yuca fields after they have planted pieces of the stem; they have harvested the tubers and carried the heavy burden back to the house; they then frequently turn over to the younger members of the household -- girls and boys -- the task of peeling the tubers and cutting them into short pieces. When there are no children available, this lot falls likewise to the women.

... to eliminate the poison the manioc is subjecte to several processes ...

The woman sits with her feet stretched out straight ... She
sets the grate on her legs ..., the forward end pressed
against her abdomen. She takes up two large manioc
tubers ... and scrapes vigorously with alternate
movements, with the hardest pressure on the forward
stroke ... When worn down to a size where it cannot
conveniently be scraped any more, the tuber is thrown
aside, and two others are taken up ... A substantial part
of the poisonous liquid ... has been removed in the
scraping on the grater; the mash is further dried by
pounding it through a tightly woven basket-sieve.

Irving Goldman, *The Cubeo: Indians of the
Northwest Amazon* (1936)

*Kubeo women grating cassava
Rio Kuduyarí, Vaupés*

When the tubers are peeled and cut into small pieces, they must be
grated to form a kind of mash which will then be subjected to a process
to extract the poisonous water in which they have been allowed to set
all night. This work is hard, requiring considerable muscular strength.
During this session, however, the women often engage in gossip and
other discussions: it is turned into a social time.

The graters are made of wood on one surface of which small
pieces of quartz have been set in a thick layer of resin which, when
thoroughly dried, holds the mineral "teeth" tight.

... scrapes vigorously with alternate movements ...

A substantial part of the poisonous liquid of the manioc has been removed in the scraping on the grater; the mash is further dried by pounding it through a tightly woven basket sieve. A tripod frame is set up and the sieve set on it at a woman's waist level ... With both fists clenched, the woman attacks the mash ...

Irving Goldman, *The Cubeo: Indians of the Northwest Amazon* (1936)

Taiwano woman crushing cassava mash
Rio Kananarí, Vaupés

The very strenuous work of squeezing, crushing and pounding the grated cassava removes much of the water in which the toxic prussic acid is dissolved. When as much of the liquid as can be expressed in this manner has been squeezed out, the mash is ready to be put into the cylindrical squeezer known as the tipitipí for further extraction of the remaining liquid.

... the woman attacks the mash ...

After the women ... had grated a sufficient quantity of Manihot, it was forcibly stuffed into an eight to nine foot long cylindrical resilient tube ... plaited out of a species of *Calathea*. The apparatus, which during the filling becomes considerably shortened and widened, was then slung by its upper loop onto one of the house beams: on the other hand a long staff was passed through the lower loop up to more than half its length, its shorter end being caught under a strong peg that had been wedged into the ground previously. Two or three women thereupon placed themselves at the longer end and forced it down with all their might, so that the yielding and shortened cylinder, owing to the pressure, gradually became longer and longer. All the watery and poisonous contents of the tubers, which the forcible stuffing had not separated as yet, were now completely expressed, collected in a large pot, thickened by long boiling and evaporation and seasoned with a strong proportion of *Capsicum*. All the poisonous constituents are volatised during the evaporation and the juice thus thickened used as sauce for meat.

> Richard Schomburgk, *Travels in British Guiana* (1922)

Kubeo women squeezing poison from yuca
Rio Kuduyarí, Vaupés

The ingenuity of native Americans in the humid tropics is clearly indicated by their discovery of how to extract a deadly poison from an otherwise edible tuberous underground source of starch -- *yuca* or *cassava*. After soaking in water overnight and slightly fermenting, the material is subjected to pressure in a wicker-like cylindrical tube, expressing much of the water with a good part of the cynogenic glycoside. This plant -- the ''bitter'' variety of *Manihot esculenta* -- is the only major food plant in the world that is utilised after the extraction of a poison. It has been suggested that the edibility of the yuca plant might have been discovered by peoples who were employing the extracted water or ''juice'' as a fish poison.

... the poisonous contents were now completely expressed ...

There are many kinds of drinks that the women know
how to make from the toasted manioc meal ... The
favorite drink, however, is kashirí, a kind of lightly
alcoholic beer.

Theodor Koch-Grünberg, *Zwei Jahre unter
den Indianern* (1910)

*Kubeo woman preparing kashirí for festival
Rio Kuduyarí, Vaupés*

Kashirí or *chicha de yuca* is prepared in enormous quantities for
dances and other festivals. For several days, the women sit and chew
fariña or yuca meal and, when it is well salivated, it is spat out into
a dug-out canoe or into ceramic pots and left to ferment. Chicha thus
prepared may be made also of numerous fruits, such as bananas,
pineapples or peach palm.

... kashiri, a kind of lightly alcoholic beer.

The traditional Cubeo attitude toward drink and intoxication is religious, not secular ... Secular drinking for them is distasteful, and it is rare to see an Indian drunk except at a ceremony ... Intoxication is a sacred state.

Irving Goldman, *The Cubeo: Indians of the Northwest Amazon* (1936)

Kubeo dancer intoxicated with chicha de yuca
Rio Kuduyarí, Vaupés

Great quantities of chicha, prepared usually from *yuca (Manihot esculenta)* or from various fruits, is normally consumed at almost all festivals in all tribes. These people view intoxication in a very different way from western cultures. Some of the tribes explain intoxication, especially that produced by drinking the hallucinogenic caapi, as the absence of the soul from the body -- that when the physical effects of the intoxication disappear, the soul has returned to the body. Intoxication amongst the Indians of the Colombian Amazon rarely leads to fighting or physical harm, unless the native is unfortunate enough to have procured from white men the highly toxic, poorly distilled aguardiente or other liquor. The locally made chichas are fermented and have a very low concentration of alcohol.

Intoxication is a sacred state.

The large saubas (leaf-cutters) and white ants are an occasional luxury when nothing else is to be had in the wet season ...

Alfred Russel Wallace, *A Narrative of Travels on the Amazon and Rio Negro* (1853)

Yukuna boys with ant traps
Caño Guacayá, Rio Miritiparaná, Amazonas

One of the pleasures that the boys enjoy is watching and taking care of these ingenious ant traps. Thousands of ants are collected and they provide a special gourmet addition to festival lunches. Not all tribes eat ants, but most of those living in the upper Rio Negro and Vaupés find them a great delicacy.

... an occasional luxury ...

We entered through a narrow pathway called an *estrada*, whose gateway was guarded by a splendid palm tree, like a Cerberus at the gates of dark Hades. The *estrada* led us past one hundred to one hundred and fifty rubber trees, as it wound its way over brooks and fallen trees.

Algot Lange, *In the Amazon Jungle* (1912)

Permanent rubber-tapping settlement
Upper Rio Vaupés, Vaupés

Along the Vaupés River, there are numerous centres of rubber tapping, almost the only industry of the whole area. The main house, where the *impresario* lives, serves as a kind of headquarters for ten or more huts, found usually on the banks of hinterland creeks, where the Indian tappers themselves live during the four or five months when rubber trees are cut. The impresario advances food, clothing and other merchandise, and receives the product of his tappers in return. Many of the impresarios build rather permanent farms and some, like the one shown in this picture, have planted rubber trees in the vicinity of their houses.

From the rubber-tappers' shacks, hidden in the forest, start the trails that lead from one rubber tree to another. These tapping circuits or *estradas* plunge boldly into the deepest forest, twisting and writhing, finally to return to their starting point. The native never clears a wide path which would make easier his daily progress through the forest to cut his trees and later to gather the latex. The mind of the Indian rebels at a wide and neatly cleaned path. Indeed, his almost imperceptible trail is to him as open and obvious as the streams and rivers on which he plies his canoe.

... guarded by a splendid palm tree ...

I have lived in the muddy swamps in the solitude of the forests with my crew of malaria-ridden men cutting the bark of the trees that have white blood like that of the gods.

José Eustásio Rivera, *La Vorágine* (1924)

Rubber tree and tapper
Rio Loretoyacu, Amazonas

Rubber trees abound throughout the Amazon Valley, but the best kind -- *Hevea brasiliensis* -- occurs almost exclusively south of the great river itself. *Hevea brasiliensis* grows only along Colombia's 65 miles of Amazon River bank. There are, consequently, probably fewer than 15,000 individuals of this kind of rubber tree in Colombian territory. In this small area, nevertheless, the trees occur densely and are rich in the white milky latex from which rubber is prepared.

During the Second World War, intensive tapping was carried out in the Leticia area in spite of the fact that improper methods employed in earlier years of this century had caused the trunks to grow rough, with great knobs and gnarls and flutings which made modern technical tapping difficult, if not impossible.

These rubber trees, locally called *seringas,* dwell happily in the lowest parts of the forest which, deeply flooded for nearly half the year, never wholly dry out. Each tapper or *seringueiro* has two circuits of about 120 trees which he bleeds on alternate days.

... white blood like that of the gods.

... the seringueiro (''rubber tapper'') leaves his hut in the morning ... Arrived on his estrada (''tapping circuit''), he proceeds to cut the trunk of each rubber tree. On the lower side of each cutting he fixes ... a small tin goblet or cup. This operation generally takes place before eleven o'clock. Towards noon, the goblets are nearly full of the viscous juice. They are then removed from the tree and their contents emptied into a pail, which the seringueiro carries to the hut. The extraction being terminated, next comes the coagulation ...

Baron de Santa-Anna Nery, *The Land of the Amazons* (1901)

Kubeo woman and her son tapping a rubber tree
Rio Tuy, Vaupés

The story of rubber tapping in the late 19th and early part of the present century is horrendous -- for many parts of the whole Amazon Valley a narrative of cheating, mistreatment, torture and even wanton murder of the Indian. The successful transplantation of the major rubber tree -- *Hevea brasiliensis* -- to Asia and its domestication in well run plantations a century ago by the British put an end to the nefarious forest industry. It accomplished two ends: a constant and adequate supply of better rubber completely changed life the world over; and the health and lives of thousands of defenseless Indians in the Amazon were saved when the procurement of rubber from wild trees died out.

When, during the Second World War, the Japanese took over most of southeast Asia, the supply of plantation rubber was drastically cut. The Amazonian countries then bent their efforts to resuscitate the production of rubber from the wild stands. The northwest Amazon in Colombia is particularly rich in rubber trees, and the Indians of that region resumed the tapping of rubber. As an emergency measure and because of distances and lack of technical knowledge easily available, tapping was primitive, not carefully controlled as in the plantations. But rubber production during these war years -- thanks to governmental and ecclesiastical supervision -- was regulated and the incredible conditions of yore did not prevail.

... the seringueiro leaves his hut in the morning ...

The thick, tough, coriaceous and very glossy leaves ...
are admirably adapted to withstand drought ... This is a
highly xerophytic plant.

Richard Evans Schultes, *Caldasia* (1944)

Dwarf rubber tree
Yapobodá, Rio Kuduyarí, Vaupés

The best rubber trees, from the point of view of commercial
exploitation, are members of the genus Hevea. There are ten species
of Hevea, but only three yield rubber of a grade high enough to be
utilised. Most of the rubber trees are forest giants, some native to
swampy sites, others growing on higher, well drained soil in the
humid tropics of the Amazon and Orinoco areas and the Guianas of
South America.

Botanists have searched the innermost recesses of the Amazon
jungles to discover new types of rubber trees and to acquire a deeper
understanding of types that have long been known. One of the most
interesting new discoveries is a dwarf variety that grows atop the
ancient, flat sandstone mountains in the Colombian Amazon.
Although yielding a good rubber, this treelet -- usually between six
and ten feet tall -- cannot be tapped because of its diminutive size. It
grows where rainfall is plentiful, but has had to adapt to drought
conditions since the sandy and stony habitat rarely retains water for
any period of time. One of the hopes of geneticists is that they may
be able to breed for plantations a lower rubber tree that will not be so
susceptible to wind damage. Use of germ plasm from this small
variety -- botanically known as *Hevea nitida* var. *toxicodendroides* --
may be extremely valuable in future research on the commercial
rubber tree of plantations, *H. brasiliensis*. This example is only one
of many that supports conservation of wild plants from wanton
extinction with the bulldozer.

... admirably adapted to withstand drought.

The *seringueiro* is free in his movements and in his mind, he is a quick and keen observer of nature and an expert in knowledge of the cries and calls of the animals of the forest. He knows their habits and hiding-places to perfection, and he could probably astonish the naturalist by informing him of many things he has observed that his brother scientist never has heard of. He knows the names of the trees and plants in the forest and what they can be used for, though his knowledge of them is often supplemented by superstitious imaginings. He knows the multitudinous fish of the Amazon, whether they are to be caught with a net, speared, or shot with a bow and arrow, or, if the hunter is of a progressive disposition, shot with rifle ball.

Lange Algot, *In the Amazon Jungle* (1912)

Tukano rubber-tapper bringing in his day's production Soratama, Rio Apaporis, Amazonas

When the tapping of the wild stands of Hevea trees in various parts of the Amazon Valley was resuscitated during the Second World War when the Japanese occupation of southeastern Asia cut off the supply of rubber from the plantations, the Indians and other inhabitants had to learn over again how to exploit the trees of their forests. With governmental assistance, they contributed a most important commodity for the war effort. Several generations had gone by since the demise of the once nefarious forest industry, and descendants of the once mistreated rubber-tappers, even though they remembered the suffering of their forefathers, demonstrated their adaptability in the new industry. The rubber-tapper of the 1940's and 1950's was spared the ignominious and cruel exploitation so common in many parts of the Amazon in the early part of the century, and he responded in kind. While rubber-tapping in the forest was still a difficult type of work, it made possible for some to save and acquire some of the amenities of life that they had hitherto not known.

... free in his movements and in his mind ...

A *seringueiro* "rubber tapper" had collected his product, and when I went to the smoking-hut I found him busy turning over and over a big stick resting on two horizontal guides built on both sides of a funnel from which a dense smoke was issuing. On the middle of the stick was huge ball of rubber. Over this he kept pouring the milk from a tin-basin. Gradually the substance lost its liquidity and coagulated into a beautiful yellow-brown mass which was rubber in its first crude shipping state.

Algot Lange, *In the Amazon Jungle* (1912)

Coagulation of rubber latex
Rio Loretoyacu, Amazonas

During the resuscitation of the wild rubber-producing industry during and after the Second World War, most rubber-tappers in Brazil and Peru continued the old method of coagulating the latex into balls of rubber. This method was followed in the Leticia area, where Brazilian influence was strong and most of the tappers were Brazilians. In most of the Colombian Amazon, a more modern method -- coagulating latex with acids and pressing it into sheets -- was taught and adopted. The older method was time-consuming and detrimental to health, the smoke causing in many cases severe conjunctivitis. The old method, however, was vehemently defended by rubber-tappers in Brazil, despite the advantages to health and the lessening of hours of work demanded by the acid-coagulation method.

... the milk coagulated into a beautiful mass which was rubber ...

The fire lighted, the seringuiero (''rubber tapper'') takes a sort of palette or large wooden spatula, which he dips several times into the pail, where the sap, which is soon to become the India-rubber of commerce, has the appearance of thick cream. He then exposes his mould to the action of the column of smoke ... The liquid part quickly evaporates and a thick layer of gum elastic forms upon the mould. The operation is repeated ... As soon as he has obtained the quantity required, he cuts the ball of caoutchouc ... gives two taps on the sides of the mould and removes the thick mass of rubber that has just coagulated and exposes it to the sun, where it acquires the blackish colour it has when placed upon the market.

Baron de Santa-Anna Nery, *The Land of the Amazons* (1901)

Ball of smoked rubber ready for sale
Rio Loretoyacu, Amazonas

When the production of rubber from wild trees was resumed during the emergency of the Second World War, modern methods of coagulating the latex with acids and the preparation of thin sheets of rubber were taught. Most Colombian rubber was produced in this form. Where the Brazilian influence was felt rubber tappers choose to follow the time-honoured system of coagulating latex with smoke.

... the sap has the appearance of thick cream ...

Close to the chief's platform is suspended the "manguare," a kind of telegraph used by the Indians in conversing with others at distances from six to twelve miles. This is made possible by means of a recognized system of beats. The manguare ... consists of the hollowed out sections of two trees, one being much larger than the other. The small one is about four feet long and the other about five and a half. The small one is called the male and the large the female ... The trunks are suspended with danta hide [tapir hide] from the main cross beam ... The manguares are separated from each other by about ... two feet. Repeated taps are made upon them concurrently and independently in recognized codes. The signals are made with a heavy kind of drum-stick, having congealed crude rubber upon one end.

Joseph F. Woodroffe, *The Upper Reaches of the Amazon* (1914)

Witoto Indian sending messages on the manguare
La Chorrera, Rio Igaraparaná, Amazonas

A number of tribes -- especially the Witotos and Boras -- utilise a most ingenious invention: the *manguare*. By pounding this hollow log in different rhythms, they are able to send discreet messages for long distances. The sound travels for great distances, especially over the water of the rivers. The manguare is used for many messages, but particularly for calling tribal conferences or announcing festivals. The invention of this apparatus is lost in the dim past: the natives say only that it was given to them by the "ancient ones". Many of the individual manguares are very old, having been passed down from father to son for generations.

... a kind of telegraph ...

The dancers stride out side by side, holding in their right hands sticks which they thump on the ground in time. These thumping-sticks are hollow cylinders of the guarumo tree ... the surface of which is covered with many-coloured patterns.

Theodor Koch-Grünberg, *Zwei Jahre unter den Indianern* (1910)

Makuna dancers with thumping sticks
Rio Piraparaná, Vaupés

One of the curious musical instruments of the Indians of many parts of the Colombian Amazon is the thumping stick. It is a hollowed out trunk of young guarumo trees (Cecropia species), beautifully painted. When these are rhythmically stamped with each footstep, the sound travels far and wide over the rivers and lends an air of reality to the music and accompanies the thuds and the singing.

... sticks which they thump on the ground in time ...

There followed another dance with the women, more rapid than the preceding ones and this time not accompanied by songs but by panpipes.

Gerardo Reichel-Dolmatoff, *The Shaman and the Jaguar* (1975)

Makuna boy with panpipes
Rio Piraparaná, Vaupés

At a very young age the boys take to music and learn to use the panpipes. It is, in fact, usually the younger members of the tribe who pipe the music for the numerous dances. They eagerly accept the opportunity of participating in the rituals and dances to which normally they are not actually a part, and they do it well and with pride. The seriousness with which they make their contribution to tribal celebrations is usually acknowledged by the leader of the ceremony or dance in the monotonous chanting with which he closes the event.

There followed another dance ... accompanied by panpipes ...

Dances may be of a profane character and may be performed only for amusement; but, like chants, they very frequently have a magico-religious efficacy and are executed to influence the supernatural beings or to coerce Nature.

Alfred Métraux's ''Religion and Shamanism''
in J. H. Steward (Ed.), *Handbook of South
American Indians* Bureau of Ethnology
Bulletin (1949)

Kai-ya-ree dancers amongst the Yukunas
Rio Miritiparaná, Amazonas

The Kai-ya-ree, which the Yukunas celebrate in April at the time of the harvest of the peach-palm or *chontaduro (Guilielma speciosa)* appears to have as its main purpose the propitiation of evils spirits by a reenactment in dance pageantry of the supposed evolutionary history of the tribe. Lasting a full 48 hours, it comprises a series of at least 30 separate dances, each with a different mask representing usually an animal. In the accompanying photograph may be seen a group of dancers wearing the round, human-like mask epitomising the spirit of the devil. In the centre, a dancer is wearing the ant-eater mask, depicting the long, arcuate snout of this animal.

... to influence the supernatural beings or coerce Nature.

Hey hey hey, ya hey hey hey
Cha wee nai yo, cha wee nai yo o
Hey hey hey, ya hey hey hey
Ya wee nai yo, ya wee nai yo
Ya wee nai yo, ya wee nai yo
E ya ree ya ka ba a, ee ya ka
Cha wee nai yo o
Hey hey hey
E ya ka
Cha wee nai yo o
Hey hey hey, ya hey hey hey

Yakuna song for Dance of the Devil
(Cha Wee Nai Yo)

Yukuna dancer with devil mask
Caño Guacayá, Rio Miritiparaná, Amazonas

The first and longest dance in the three- or four-day Kai-ya-ree Dance is dedicated to the ''devil''or the force of evil or darkness. It is accompanied by singing on the part of all the men who participate in this part of the dance. The monotonous, pentatonic but beautiful song coming out through the masks creates a truly weird and convincingly unworldly atmosphere in which the long series of dances is born.

Cha wee nai yo …

These Indians find a pretext for dancing in everything. A dance during preparations for war or when the fighters come back in victory; a dance to entertain a chieftain or to mourn a death; a dance to celebrate the harvest of the tapioca root or the ripening of the peach-palm, the *inga* or the *umarí;* a dance to observe a girl's puberty or the end of communal work.

Gastão Cruls, *Hileia Amazónica* (1944)

*Yukunas dressed for the butterfly dance
Rio Miritiparaná, Amazonas.*

One of the ritualistic dances in the Kai-ya-ree Dance celebrated by the Yukunas in April when the peach-palm fruit is ripe pays honour to the large, strikingly iridescent blue morpho butterfly that flits aimlessly and lackadaisically along trails in the Amazon jungles. The masks, made of balsa wood and decorated with paints prepared from coloured earths and plant dyes, represent the butterfly. The dance is slow, stately but lazily aimless and, like the morpho himself, never fails to thrill the onlooker with wonder and admiration that such grace of movement could be found in such an asperous environment.

... a pretext for dancing in everything.

The cylindrical head-mask together with its "ears" are made of very light wood painted with many colours. The covering of the head and the jacket and sleeves are made of reddish tururí-bark.

Theodor Koch-Grünberg, *Zwei Jahre unter den Indianern* (1910)

Yukuna dancers with sun-masks
Caño Guacayá, Rio Miritiparaná, Amazonas

According to several Yukuna informants, the Kai-ya-ree Dance is performed for many purposes, particularly to pacify the forces of darkness. One of the many aspects of the four-day ceremony involves a beautiful dance honouring the sun. The wooden mask, made of balsa wood and painted with bright colours, is worn during this dance that is believed to thwart the wiles of the spirits of darkness. A yellow face representing the sun is painted on the cylindrical mask.

... a beautiful dance honouring the sun.

The role of the shamans and elders in defining and explaining these graphic symbols and their associations is decisive. Their explanations, paired with admonitions and couched in the language of myth and ritual, tend to form a consensus nobody would put in doubt ... These symbols ... must be viewed within their specific social context and can be understood only within the framework of the cosmological and ecological concepts that underlie the native culture. They have no ''magical'' or ''religious'' basis; the supernatural sphere comes into play only because it effectively reinforces the message. By thus being sanctioned the graphic symbol acquires an almost threatening character and in this manner its message is more likely to be heeded.

Gerardo Reichel-Dolmatoff, *Beyond the Milky Way* (1978)

Painted Indian house
Rio Piraparaná, Vaupés

Some tribes in Colombia's Amazonian forest regularly paint the outer walls of their thatch-and-bark houses. There is a great diversity of motifs, but some of the figures seem to be copied from the ancient petroglyphs which, many centuries ago, were graven on the rocks of the many waterfalls and rapids of the rivers. Anthropologists have attempted to explain a number of the figures without any great success.

... couched in the language of myth and ritual ...

On the front of many bark-faced *malocas*, I have seen decorations of men, animals, tools of everyday use or patterns of weaving more or less characteristically drawn in black. Sometimes, however, the walls of the house have much richer ornamentation in variegated colours -- black, red, yellow -- in an orderly artistic arrangement.

> Theodor Koch-Grünberg, *Zwei Jahre unter den Indianern* (1910)

Makuna house-painting
Rio Popeyacá, Amazonas

Whereas most of the painted malocas that still exist in the Vaupés are intricately decorated with geometric designs, supernaturally seen beings from caapi-intoxications and other designs not daily encountered, some of the bark walls are painted in carbon-black with prosaic figures -- men, animals, tools of everyday use and the patterns of weaving. Whether or not this difference from the more magically and supernaturally oriented paintings is significant or due to lack of time or care is not known, but both are of interest and are characteristic of the region.

… decorations of men … characteristically drawn in black.

The necklaces are matters of importance, for they disclose the status of the wearers. The skill of a warrior as a hunter, his bravery in war, is proved by the character of the teeth that circle his neck: the more successful the hunter, the finer the teeth he wears, the more numerous the adornments of his family.

Thomas Whiffen, *The Northwest Amazons*
(1915)

Kofán chieftain with boar's tooth necklace
Rio Sucumbios, Putumayo

Nowhere in Colombia's Amazonia is the use of elaborate necklaces of animal teeth so characteristic as amongst the Kofáns of the Putumayo. Chieftains, leaders, medicine-men and curare-makers wear strings of teeth, usually of the wild boar so abundant in their forests, as adornments of office. Similarly, brilliant feathers of the macaw, worn in plugs inserted through the ear-lobes or the septum of the nose serve as badges of office. In addition, the Kofáns like to wear strings of tiny, brightly-coloured glass beads won by trade. Many yards of these beads or *chaquira,* strung on durable threads of cumare fibre are permanently wound about the neck. Sometimes pounds of these beads form a fantastic collar, the size and weight of which indicate the wearer's importance in his tribe.

... the more successful the hunter, the finer the teeth he wears ...

My Indians were excessively proud of their own artistic products ... The charming humour that is evident in many of their drawings attests to their delight in creative work.

> Theodor Koch-Grünberg, *Zwei Jahre unter den Indianern* (1909)

Kubeos making balsa thumping-sticks
Rio Vaupés, Vaupés

One of the typical musical instruments of the Kubeo and Tukano peoples in the Vaupés is the curious thumping-stick. This is made from small balsa trunks the pith of which is hollowed out with fire and decorated on the outside either by burning or by earth-painting. Decorations usually consist of geometric designs at the base and a stereotyped animal on the upper part of the stick. These sticks are thumped rhythmically during many of the dances and their resonance helps wonderfully in enlivening the dance and marking time of the steps.

... delight in creative work.

All those things in which the Indian takes great pains and with which he adorns himself are delicate, soft, symmetrical and worked with a feminine touch quite in contrast to the roughness of the environment.

Enrique Pérez-Arbeláez, *Vaupés* (1950)

Tukano featherwork
Yutica, Rio Vaupés

The Indian's most valuable material possession is his box of feather dancing ornaments. These he makes with all the care and love of which he is capable. He packs them in a low, rectangular box made from leaves. He hangs the box in a high place under the rafters, where just enough smoke will reach it to coat it with an insect- and mould-repelling layer of pungent resins. Several times a year, the box is opened and the featherwork is aired in preparation for wearing. And when tribesmen from neighbouring areas come together for festivals, there are often excited contests for the best feather-dressing. In the Vaupés, the Tukanos are recognised as the masters of featherwork.

... delicate ... in contrast to the roughness.

Some ignorant travellers and colonists call these Indians a lazy race. Man in general will not be active without an object. Now when the Indian has caught plenty of fish and killed game enough to last him for a week, what need has he to range the forest? He has no idea of making pleasure grounds. Money is of no use to him, for in these wilds there are no markets for him to frequent nor milliners' shops for his wife and daughters; he has no taxes to pay, no highways to keep up; he lies in his hammock both night and day ..., and in it he forms his bow, makes his arrows and repairs his fishing tackle. But as soon as he has consumed his provisions, he then rouses himself and, like a lion, scours the forest in quest of food.

Charles Waterton, *Wanderings in South America* (1825)

Taiwano opening dugout canoe
Rio Kananarí, Vaupés

The critical point in the construction of the ever-present dugout canoe is the firing of the hollowed-out trunk to open or widen it. If this is not done with extreme care, the trunk will split, and days or weeks of work are lost. The Indians who do this critical work must keep to a strict diet the day before the task: no chili-pepper, for example, must be eaten, else the canoe will surely split! The greatest of care must be taken in wedging in the cross pieces which prevent the cooling trunk from shrinking and it is not uncommon to see five or six expert canoe-builders in heated conference as to just how wide a bar can safely be wedged into the hollowed-out trunk. The building of a canoe usually requires from seven to ten days.

... will not be active without an object.

That evening was a beautiful one, and we made camp at an old *mitasava* [sic], the Indian traveller's staging point, where he erects a small palm-thatch shelter.

Brian Moser and Donald Taylor, *The Cocaine Eaters* (1965)

A miritisaba in the forest
Rio Kuduyarí, Vaupés

Along riverbanks and in the forests of the northwest Amazon, one of the most heartening apparitions is the finding of a *miritisaba* -- a temporary palm roof under which the wayfarer may defend himself overnight from the usual heavy rainfall. The natives erect these makeshift roofs for themselves to pass the night on a long trek; but they leave them in shape for anyone who may follow during the ensuing several weeks. These shelters constitute a really happy discovery by the tired traveller who, overtaken by oncoming twilight, is too weary to unpack and erect his own shanty.

... the Indian traveller's staging point ...

He will pass through small streams, lakes and swamps, and everywhere around him will stretch out an illimitable waste of waters, but all covered with a lofty virgin forest. For days, he will travel through this forest, scraping against tree-trunks, and stooping to pass beneath the leaves of prickly palms, now level with the water, though raised on stems forty feet high. In this trackless maze, the Indian finds his way with unerring certainty, and by slight indications of broken twigs or scraped bark, goes on day by day as if travelling on a beaten road.

Alfred Russel Wallace, *A Narrative of Travels on the Amazon and Rio Negro* (1889)

Foot bridge across a stream
Rio Igaraparaná, Amazonas

Water is everywhere in the Colombia Amazonian realm. Rivers, brooks and streams cut the forests into a dizzy patchwork, and, between the patches, countless rills and freshets continue the dissection of the land.

Crossing the numberless watercourses, both large and small, presents a problem to whoever, like the rubber trapper, must leave his canoe and walk through the undergrowth. It is often merely a felled trunk, slippery and moss-green, that serves to span a rill or brook. Some of the larger streams, given to unbelievable and almost momentary rises and falls in response to the heavy but capricious rains, must be more stoutly bridged, and logs and trunks are then lashed together with the strong aerial roots of the epiphytic aroids to hold firm against the force of the swells.

In this trackless maze …

The great river and its tributaries are the only means of travel through the primeval forest; a thousand yards or less from these flowing roads and the wayfarer is hopelessly lost in the overwhelming exuberance of tropical plant life, a land of grey-green twilight, where no landmark, no horizon, often not even a glimpse of the sky enables him to guess at his direction.

> Christopher Sandeman, *A Wanderer in Inca Land* (1948)

Paddling home in the peace of early evening
Rio Kubiyú, Vaupés

Placid and silent, these tortuous tributaries invite the tired hunter or fisherman to glide along their glassy surface to home and food and family -- the end of a perfect day. They provide the only avenues in this region of dense forest, and the natives hold them in deep respect. In fact, many Indians hold them as ''friends and allies of the overpowering forest''.

... the only means of travel through the forest ...

We shot the foaming billows, and fortunately reached the edge of the irrestibly seething caldron, but not without our boat being filled with water near to sinking ...

Richard Schomburgk, *Travels in British Guiana* (1922)

Barasana navigators bailing out
Rio Piraparaná, Vaupés

With an almost amphibious spirit, the Barasana Indians traffic up and down their rapid-choked Piraparaná or ''river of fishes''. Approaching a rapid, their zest for the sport of pitting eye, mind and might against the water heightens. They literally ''shoot the rapids'', swiftly changing their course with a few deft twists of the paddles in the very midst of the churning waters to avoid disaster. Usually, so great is their skill, the dugouts come through as dry as when they entered the cauldron. But not always, for often canoes and navigators glide to calmness below a rapid thoroughly watered down. Nothing daunted, they dry themselves in the sun and bail out the canoe, ready for the next adventure. Seldom does an Indian lose his life in these rapids.

... filled with water near to sinking.

.. the passage being generally in the middle of the river, among rocks, where the water rushes furiously. The falls were not more than three or four feet each; but, to pull a loaded canoe up these, against the foaming waters of a large river, was a matter of the greatest difficulty for my dozen Indians, their only resting place being often breast deep in water, where it was a matter of wonder that they could stand against the current, much less exert any force to pull the canoe.

Alfred R. Wallace, *A Narrative of Travels on the Amazon and Rio Negro* (1853)

Passing the rapids of Tatú
Rio Vaupés, Vaupés

The Indians who navigate in the rapid-choked rivers of the Colombian Amazon become extraordinarily skillful and come to know every rock, every passage and the very best and safest way to operate. This skill is learned from early age and it involves becoming familiar with the waters themselves that change almost daily, sometimes in hours, with the rise and fall of the rivers. Quite contrary to the general belief of visitors not familiar with the treachery of these rapids, it is normally a far easier task to navigate downstream than to guide or pull a canoe upstream through the rocks.

... a matter of the greatest difficulty ...

Like shuttles in a loom, *batalones* weave a pattern of
human life on the rivers.

Enrique Pérez-Arbeláez, *Vaupés* (1950)

A boatload of thatch
Rio Vaupés, Vaupés

Not long ago, all heavy transport on the Vaupés and other rivers of
the rubber or balata-producing areas was done in enormous and, to the
eye, unwieldy *batalones* -- huge dugouts with built-up plank
washboards. Unbelievably safe in the innumerable rapids, the
batalones are moved by a score or more of Indian paddlers. Before the
days of air transport of forest produce from the Vaupés to Bogotá,
balata and other products of Indian labour in the jungles were sent to
Manáos for sale in batalones, the round trip requiring three to three
and a half months of endless days of paddling and struggle with the
forty or more rapids between Mitú and Manáos. Although its utility
is restricted nowadays where outboard motors or small launches can
be employed, the batalon still ''weaves a pattern of human life on the
rivers''.

... a pattern of human life ...

Latterly, [the Putumayo River] has been explored and mapped, and there is much talk of a steamboat line from Pará, by this river, to New Granada. The line, no doubt, is feasible enough, but there is no commercial necessity for it, and if established, I fear it will be only another subsidized toy ...

Herbert H. Smith, *Brazil: the Amazons and the Coast* (1879)

The wood-burning side-paddler "Ciudad de Neiva"
Caucaya, Rio Putumayo, Putumayo

With the ease of regular flights today to centres like Leticia and Mitú, overland or river transportation has disappeared, at least on a commercial basis. The slow, cumbersome river boats that plied the Putumayo to Leticia were so slow that, with the advent of air transport, it was found far from economical to keep the boats running. Thus passed an epoch in the story of the Colombian Amazon -- an epoch that many will recall with nostalgia.

... only another subsidised toy ...

From the time it crosses the Guaviare until it lands on the river in Mitú, capital of the Vaupés, the aeroplane takes one hour and a half. It flies 375 miles and throbs over an immense flatness covered with rain-jungle, typical Amazonian forest cut only by the winding path of the rivers, by forgotten ox-bows and by fens where the sun sees itself mirrored.

Enrique Pérez-Arbeláez, *Vaupés* (1950)

Take-off of Catalina aeroplane
Karurú, Rio Vaupés, Vaupés

The few small settlements of civilised people along the Amazonian rivers of Colombia are dedicated to the exploitation of rubber, fibre or other products of the forest or to missionary work. The arrival of the aeroplane is always an occasion for great joy, for it is the only link that these isolated folk have with the populous centres of the country. Their few necessities and newspapers are brought in by these flying boats -- termed "wico" or big bird by the Kubeo Indians -- and their produce is taken out to the industrial centres up in the Andes mountains. Many times, too, these craft fly errands of mercy, taking young and old to centres where good medical attention can be sought.

It is of curious interest that this most primitive part of Colombia is actually to a great extent much more air-minded than many places in Europe and the United States -- and rightfully so, because the journey from Mitú to Bogotá, a major expedition of some two months by canoe and overland, is flown in a mere three and one half hours. Here, where pack animals and motor-cars are virtually unknown, transportation is carried out by the most primitive means -- human carriers and dugout paddle-canoes -- and the most advanced -- the airship!

... a mere three and a half hours.

Acknowledgements

It is with gratitude that I thank the following colleagues and friends for permission to reproduce several of their illustrations:

Dr. José Cuatrecasas for the photograph of a tame tapir in a Kofán Indian house (page 139).

Professor Hernando García-Barriga for three photographs: the mountains and cataracts of Araracuara (page 71); a Kubeo girl (page 125); the oco-yajé *(Diplopteris Cabrerana)* (page 181).

Dr. Mariano Ospina H. for the panorama of the Valley of Sibundoy (page 47).

Dr. Timothy C. Plowman for the photograph of the Tukano Indian pounding coca leaves (page 205).

My deep appreciation goes to the expert photographer in Bogota, Mr. Frederico Birkigt, whose artistic and scientific photographic skill enlarged from my Kodak 120 negatives taken with a Rolleiflex camera the superb prints that are published in this volume.

I am grateful to Dr. Stephen Hugh-Jones, Kings College, University of Cambridge, England, for reviewing the anthropological information in the introduction.

Finally, I am indebted to my secretary, Mrs. Mary Gaudet, for her patience and understanding, editorial suggestions and help over the years of preparation of this book.

RICHARD EVANS SCHULTES

Synergetic Press

Synergetic Press focuses on biospheric publications, including:

The Biosphere Catalogue, edited by T. Parrish Snyder, a comprehensive presentation of the biosphere with contributions from over thirty leading figures in fields ranging from atmosphere, hydrosphere, geosphere, plants and animals to cultures, cities, space biospheres, genetics and travel.

Space Biospheres, by John Allen and Mark Nelson, a concise integrative model of Biosphere I (the Earth's biosphere) and introduction to Biosphere II, a project to create a two-acre enclosed and recycling ecological system.

The Biosphere, by Vladimir Vernadsky, the first English edition of his classic work, originally published in Russian in 1926.

Traces of Bygone Biospheres, by Andrey Lapo, an examination of the role of life in geological processes and in the evolution of the biosphere of the Earth.

Where the Gods Reign is the first in a series of publications on biomes of the biosphere. In addition, Synergetic Press publishes works from the vanguard in poetry, fiction and drama. Please write one of our office locations for a complete booklist.

Post Office Box 689
Oracle, Arizona 85623

SP

24 Old Gloucester Street
London WC1 3AL England